WORLD DEVELOPMENT REPORT

THE
CHANGING NATURE
OF WORK

D1341470

A World Bank Group Flagship Report

T H E
CHANGING NATURE
OF WORK

WORLD BANK GROUP

ISSN, ISBN, e-ISBN, and DOI:

Softcover
ISSN: 0163-5085
ISBN: 978-1-4648-1328-3
e-ISBN: 978-1-4648-1356-6
DOI: 10.1596/978-1-4648-1328-3

Hardcover
ISSN: 0163-5085
ISBN: 978-1-4648-1342-9
DOI: 10.1596/978-1-4648-1342-9

Cover art: Diego Rivera, *The Making of a Fresco Showing the Building of a City,* 1931, fresco, 271 by 357 inches, gift of William Gerstle. Image copyright © San Francisco Art Institute. Used with permission; further permission required for reuse.

Cover design: Weight Creative, Vancouver, British Columbia, Canada.

Interior design: Debra Naylor, Naylor Design, Inc., Washington, DC.

Contents

Foreword

At a time when the global economy is growing and the poverty rate is the lowest in recorded history, it would be easy to become complacent and overlook looming challenges. One of the most critical is the future of work, the subject of the 2019 *World Development Report*.

"Machines are coming to take our jobs" has been a concern for hundreds of years—at least since the industrialization of weaving in the early 18th century, which raised productivity and also fears that thousands of workers would be thrown out on the streets. Innovation and technological progress have caused disruption, but they have created more prosperity than they have destroyed. Yet today, we are riding a new wave of uncertainty as the pace of innovation continues to accelerate and technology affects every part of our lives.

We know that robots are taking over thousands of routine tasks and will eliminate many low-skill jobs in advanced economies and developing countries. At the same time, technology is creating opportunities, paving the way for new and altered jobs, increasing productivity, and improving the delivery of public services. When we consider the scope of the challenge to prepare for the future of work, it is important to understand that many children currently in primary school will work in jobs as adults that do not even exist today.

That is why this Report emphasizes the primacy of human capital in meeting a challenge that, by its very definition, resists simple and prescriptive solutions. Many jobs today, and many more in the near future, will require specific skills—a combination of technological know-how, problem-solving, and critical thinking—as well as soft skills such as perseverance, collaboration, and empathy. The days of staying in one job, or with one company, for decades are waning. In the gig economy, workers will likely have many gigs over the course of their careers, which means they will have to be lifelong learners.

Innovation will continue to accelerate, but developing countries will need to take rapid action to ensure they can compete in the economy of the future. They will have to invest in their people with a fierce sense of urgency—especially in health and education, which are the building blocks of human capital—to harness the benefits of technology and to blunt its worst disruptions. But right now too many countries are not making these critical investments.

Our Human Capital Project aims to fix that. This study unveils our new Human Capital Index, which measures the consequences of neglecting investments in human capital in terms of the lost productivity of the next generation of workers. In countries with the lowest human capital investments today, our analysis suggests that the workforce of the future will only be one-third to one-half as productive as it could be if people enjoyed full health and received a high-quality education.

Adjusting to the changing nature of work also requires rethinking the social contract. We need new ways to invest in people and to protect them, regardless of their employment status. Yet four out of five people in developing countries have never known what it means to live with social protection. With 2 billion people already working in the informal sector—unprotected by stable wage employment, social safety nets, or the benefits of education—new working patterns are adding to a dilemma that predates the latest innovations.

This Report challenges governments to take better care of their citizens and calls for a universal, guaranteed minimum level of social protection. It can be done with the right reforms, such as ending unhelpful subsidies; improving labor market regulations; and, globally, overhauling taxation policies. Investing in human capital is not just a concern for ministers of health and education; it should also be a top priority for heads of state and ministers of finance. The Human Capital Project will put the evidence squarely in front of those decision makers, and the index will make it hard to ignore.

The 2019 *World Development Report* is unique in its transparency. For the first time since the World Bank began publishing the WDR in 1978, we made an updated draft publicly available, online each week, throughout the writing process. For over seven months, it has benefited from thousands of comments and ideas from development practitioners, government officials, scholars, and readers from all over the world. I hope many of you will have already read the Report. Over 400,000 downloads later (and counting), I am pleased to present it to you in its final form.

Jim Yong Kim
President
The World Bank Group

Overview

There has never been a time when mankind was not afraid of where its talent for innovation might lead. In the 19th century, Karl Marx worried that "machinery does not just act as a superior competitor to the worker, always on the point of making him superfluous. It is the most powerful weapon for suppressing strikes."[1] John Maynard Keynes warned in 1930 of widespread unemployment arising from technology.[2] And yet innovation has transformed living standards. Life expectancy has gone up; basic health care and education are widespread; and most people have seen their incomes rise.

Three-quarters of the citizens of the European Union, the world's lifestyle superpower, believe that the workplace benefits from technology, according to a recent Eurobarometer survey. Two-thirds said technology will benefit society and improve their quality of life even further (figure O.1).

Despite this optimism, concerns about the future remain. People living in advanced economies are anxious about the sweeping impact of technology on employment. They hold a view that rising inequality, compounded by the advent of the gig economy (in which organizations contract with independent workers for short-term engagements), is encouraging a race to the bottom in working conditions.

This troubling scenario, however, is on balance unfounded. It is true that in some advanced economies and middle-income countries manufacturing jobs are being lost to automation. Workers undertaking routine tasks that are "codifiable" are the most vulnerable to replacement. And yet technology provides opportunities to create new jobs, increase productivity, and deliver effective public services. Through innovation, technology generates new sectors and new tasks.

FIGURE O.1 Survey respondents believe technology is improving the European economy, society, and quality of life

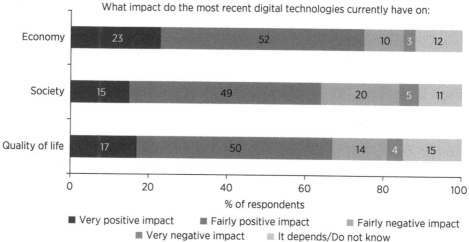

What impact do the most recent digital technologies currently have on:

	Very positive impact	Fairly positive impact	Fairly negative impact	Very negative impact	It depends/Do not know
Economy	23	52	10	3	12
Society	15	49	20	5	11
Quality of life	17	50	14	4	15

% of respondents

■ Very positive impact   ■ Fairly positive impact   ■ Fairly negative impact
■ Very negative impact   ■ It depends/Do not know

Source: WDR 2019 team, based on Special Eurobarometer 460, "Attitudes towards the Impact of Digitization and Automation on Daily Life," Question 1, European Commission, 2017.

FIGURE O.2 Recent technological advances accelerate the growth of firms

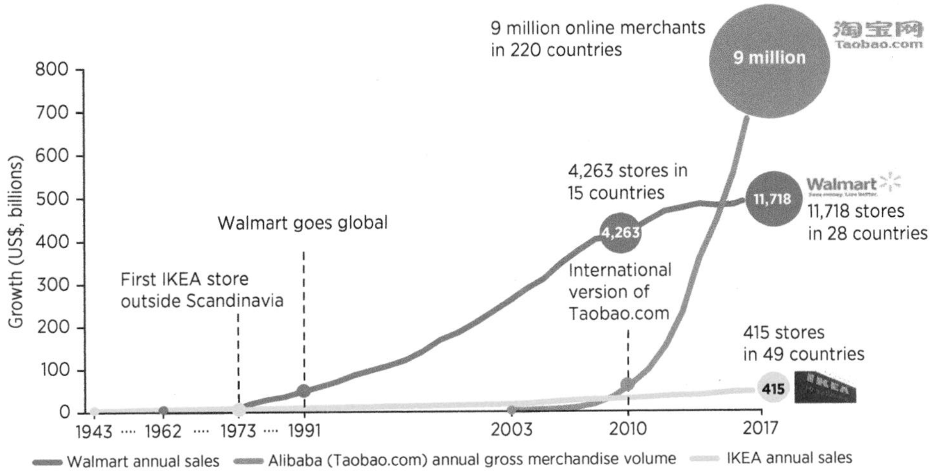

Source: WDR 2019 team, based on Walmart annual reports; Statista.com; IKEA.com; NetEase.com.

Some features of the current wave of technological progress are notable. Digital technologies allow firms to scale up or down quickly, blurring the boundaries of firms and challenging traditional production patterns. New business models—digital platform firms—are evolving from local start-ups to global behemoths, often with few employees or tangible assets (figure O.2). This new industrial organization poses policy questions in the areas of privacy, competition, and taxation. The ability of governments to raise revenues is curtailed by the virtual nature of productive assets.

The rise of platform marketplaces allows the effects of technology to reach more people more quickly than ever before. Individuals and firms need only a broadband connection to trade goods and services on online platforms. This "scale without mass" brings economic opportunity to millions of people who do not live in industrialized countries or even industrial areas.[3] The changing demand for skills reaches the same people. Automation raises the premium on high-order cognitive skills in advanced and emerging economies.

Investing in human capital is the priority to make the most of this evolving economic opportunity. Three types of skills are increasingly important in labor markets: advanced cognitive skills such as complex problem-solving, sociobehavioral skills such as teamwork, and skill combinations that are predictive of adaptability such as reasoning and self-efficacy. Building these skills requires strong human capital foundations and lifelong learning.

The foundations of human capital, created in early childhood, have thus become more important. Yet governments in developing countries do not give priority to early childhood development, and the human capital outcomes of basic schooling are suboptimal. The World Bank's new human capital index, presented in this study for the first time, highlights the link between investments in health and education and the productivity of future workers. For example, climbing from the 25th to the 75th percentile on the index brings an additional 1.4 percent annual growth over 50 years.

Creating formal jobs is the first-best policy, consistent with the International Labour Organization's decent work agenda, to seize the benefits of technological change. In many developing countries, most workers remain in low-productivity employment, often in the informal sector with little access to technology. Lack of quality private sector jobs leaves talented young people with few pathways to wage employment. High-skill university graduates currently make up almost 30 percent of the unemployed pool of labor in the Middle East and North Africa. Better adult learning opportunities enable those who have left school to reskill according to changing labor market demands.

Investments in infrastructure are also needed. Most obvious are investments in affordable access to the Internet for people in developing countries who remain unconnected. Equally important are more investments in the road, port, and municipal infrastructure on which firms, governments, and individuals rely to exploit technologies to their full potential.

Adjusting to the next wave of jobs requires social protection. Eight in 10 people in developing countries receive no social assistance, and 6 in 10 work informally without insurance.

Even in advanced economies, the payroll-based insurance model is increasingly challenged by working arrangements outside standard employment contracts. What are some new ways of protecting people? A societal minimum that provides support independent of employment is one option. This model, which would include mandated and voluntary social insurance, could reach many more people.

Social protection can be strengthened by expanding overall coverage that prioritizes the neediest people in society. Placing community health workers on the government's payroll is a step in the right direction. A universal basic income is another possibility, but it is untested and fiscally prohibitive for emerging economies. Enhanced social assistance and insurance systems would reduce the burden of risk management on labor regulation. As people become better protected through such systems, labor regulation could, where appropriate, be made more balanced to facilitate movement between jobs.

For societies to benefit from the potential that technology offers, they would need a new social contract centered on larger investments in human capital and progressively provided universal social protection (figure O.3). However, social inclusion requires fiscal space, and many developing countries lack the finances because of inadequate tax bases, large informal sectors, and inefficient administration.

And yet there is plenty of room for improvement through, for example, better collection of property taxes in urban municipalities or the introduction of excise taxes on sugar or tobacco. The latter would have direct health benefits as well. Levying indirect taxes, reforming subsidies, and reducing

FIGURE O.3 Responding to the changing nature of work

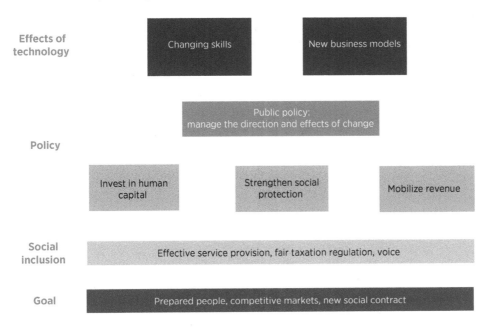

Source: WDR 2019 team.

tax avoidance by global corporations, especially among the new platform companies, are other possible sources of financing. In fact, the traditional structure of the global tax order provides opportunities for multinational corporations to engage in base erosion and profit shifting—that is, some firms allocate more profits to affiliates located in zero- or low-tax countries no matter how little business is conducted there. By some estimates, on average, 50 percent of the total foreign income of multinationals is reported in jurisdictions with an effective tax rate of less than 5 percent.[4]

Emerging economies are in the middle of a technological shift that is bringing change to the nature of work. Whatever the future holds, investment in human capital is a no-regrets policy that prepares people for the challenges ahead.

Changes in the nature of work

Several stylized facts have dominated the discussion on the changing nature of work. However, only some of them are accurate in the context of emerging economies.

First, technology is blurring the boundaries of the firm, as evident in the rise of platform marketplaces. Using digital technologies, entrepreneurs are creating global platform–based businesses that differ from the traditional production process in which inputs are provided at one end and output delivered at the other. Platform companies often generate value by creating a network effect that connects customers, producers, and providers, while facilitating interactions in a multisided model.

Compared with traditional companies, digital platforms scale up faster and at lower cost. IKEA, the Swedish company founded in 1943, waited almost 30 years before it began expanding within Europe. After more than seven decades, it achieved global annual sales revenue of US$42 billion. Using digital technology, the Chinese conglomerate Alibaba was able to reach 1 million users in two years and accumulate more than 9 million online merchants and annual sales of $700 billion in 15 years. Meanwhile, platform-based businesses are on the rise in every country—such as Flipkart in India and Jumia in Nigeria. Globally, however, integrated virtual marketplaces are posing new policy challenges in the fields of privacy, competition, and taxation.

Second, technology is reshaping the skills needed for work. The demand for less advanced skills that can be replaced by technology is declining. At the same time, the demand for advanced cognitive skills, sociobehavioral skills, and skill combinations associated with greater adaptability is rising. Already evident in developed countries, this pattern is starting to emerge in some developing countries as well. In Bolivia, the share of employment in high-skill occupations increased by 8 percentage points from 2000 to 2014. In Ethiopia, this increase was 13 percentage points. These changes show up not just through new jobs replacing old jobs, but also through the changing skills profiles of existing jobs.

Third, the idea of robots replacing workers is striking a nerve. However, the threat to jobs from technology is exaggerated—and history has repeatedly taught this lesson. The data on global industrial jobs simply do not bear out these concerns. Advanced economies have shed industrial jobs, but the rise of the industrial sector in East Asia has more than compensated for this loss (figure O.4).

The decline in industrial employment in many high-income economies over the last two decades is a well-studied trend. Portugal, Singapore, and Spain are among the countries in which the share of industrial employment has dropped 10 percent or more since 1991. This change reflects a shift in employment from manufacturing to services. By contrast, the share of industrial employment, primarily manufacturing, has remained stable in the rest of the world. In low-income countries, the proportion of the total labor force working in industry from 1991 to 2017 was consistently around 10 percent. The situation was stable in upper-middle-income countries as well, at around 23 percent. Lower-middle-income countries experienced an increase in the proportion of the labor force in the industrial sector over the same period, from 16 percent in 1991 to 19 percent in 2017. This increase may stem from the interplay of open trade and rising incomes, which generates more demand for goods, services, and technology.

In some developing countries, the share of industrial employment overall is going up. For example, in Vietnam it rose from 9 percent in 1991 to 25 percent in 2017. In the Lao People's Democratic Republic, the share of

FIGURE O.4 **Industrial jobs are falling in the West and rising in the East, but the total labor force has been increasing across the globe**

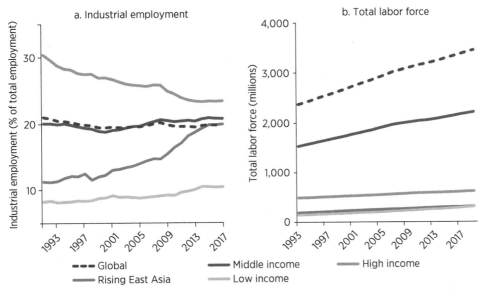

Source: WDR 2019 team, based on World Bank's World Development Indicators (database).

Note: "Rising East Asia" includes Cambodia, Indonesia, the Lao People's Democratic Republic, Mongolia, Myanmar, the Philippines, Thailand, and Vietnam.

industrial employment rose from 3 percent to 10 percent over the same period. These countries have upgraded their human capital, bringing highly skilled young workers into the labor market, who, together with new technology, upgrade manufacturing production. As a result, industrial employment in East Asia continues to rise, whereas in other developing economies it is stable.

Two forces are increasing the demand for industrial products and therefore the demand for labor in the industrial sector. On the one hand, the falling costs of connectivity are leading to more capital-intensive exports from advanced economies and more labor-intensive exports from emerging economies. On the other, rising incomes are increasing consumption of existing products and the demand for new ones.

Fourth, in many developing countries a large number of workers remain in low-productivity jobs, often in informal sector firms whose access to technology is poor. Informality has remained high over the last two decades despite improvements in the business regulatory environment (figure O.5). Indeed, the share of informal workers is as high as 90 percent in some emerging economies. Overall, about two-thirds of the labor force in these economies is informal. Informality has remained remarkably stable notwithstanding economic growth or the changing nature of work. For example, in Peru, despite all the attention focused on the issue, informality has remained constant at about 75 percent over the last 30 years. In Sub-Saharan Africa, informality remained, on average, at around 75 percent

FIGURE O.5 Informality persists in most emerging economies despite improvements in the regulatory environment

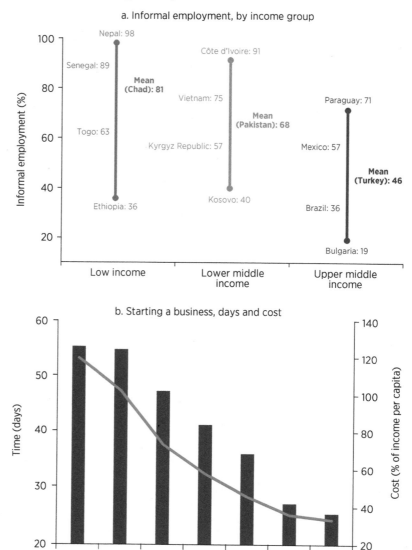

a. Informal employment, by income group

b. Starting a business, days and cost

Sources: WDR 2019 team, using household and labor force survey data from the World Bank's International Income Distribution Data Set (panel a); Djankov et al. (2002); World Bank's Doing Business Indicators (panel b).

Note: Panel a presents the latest available estimates of the shares of informal employment in emerging economies. In the sample, a person is identified as an informal worker if that person does not have a contract, social security, and health insurance and is not a member of a labor union. The sample in panel a consists of 68 emerging economies, all classified as low and middle income. Panel b shows the estimated time and cost of starting a business in 103 emerging economies.

of total employment from 2000 to 2016. In South Asia, it increased from an average of 50 percent in the 2000s to 60 percent over the period 2010–16. Addressing informality and the absence of social protection for workers continues to be the most pressing concern for emerging economies.

Fifth, technology, in particular social media, affects the *perception* of rising inequality in many countries. People have always aspired toward a higher quality of life and participation in the economic growth they see around them. Increased exposure through social media and other digital communications to different, often divergent lifestyles and opportunities only heightens this feeling. Where aspirations are linked to opportunities, the conditions are ripe for inclusive, sustainable economic growth. But if there is inequality of opportunity or a mismatch between available jobs and skills, frustration can lead to migration or societal fragmentation. The refugee crises in Europe, the war-pushed migrants from the Syrian Arab Republic, and the Arab Spring are notable manifestations of this perception.

This perception is not corroborated, however, by the data on income inequality in developing countries. Inequality in most emerging economies has declined or remained unchanged over the last decade. From 2007 to 2015, 37 of 41 of these economies experienced a decline or no change in inequality, as measured by the Gini coefficient. The four emerging economies in which inequality rose were Armenia, Bulgaria, Cameroon, and Turkey. In the Russian Federation, between 2007 and 2015 the Gini measure of inequality fell from 42 to 38. Between 2008 and 2015, the share of income of the top 10 percent of the population (based on pretax income) fell from 52 to 46 percent. The share of employment in small firms rose over that period, which improved wages relative to those of large firms.

Yet there is little to celebrate in the fact that income inequality is not, despite perceptions, rising—and even less when considering that globally 2 billion people are working in the informal economy, where so many lack any protection. Social insurance is virtually nonexistent in low-income countries, and even in upper-middle-income countries it reaches only 28 percent of the poorest people.

What can governments do?

The analysis suggests areas in which governments could act:

- Investing in human capital, particularly early childhood education, to develop high-order cognitive and sociobehavioral skills in addition to foundational skills.

- Enhancing social protection. A solid guaranteed social minimum and strengthened social insurance, complemented by reforms in labor market rules in some emerging economies, would achieve this goal.

- Creating fiscal space for public financing of human capital development and social protection. Property taxes in large cities, excise taxes on sugar or tobacco, and carbon taxes are among the ways to increase a government's revenue. Another is to eliminate the tax avoidance techniques that many firms use to increase their profits. Governments can optimize their taxation policy and improve tax administration to increase revenue without resorting to tax rate increases.

The most significant investments that people, firms, and governments can make in the changing nature of work are in enhancing human capital. A basic level of human capital, such as literacy and numeracy, is needed for economic survival. The growing role of technology in life and business means that all types of jobs (including low-skill ones) require more advanced cognitive skills. The role of human capital is also enhanced because of the rising demand for sociobehavioral skills. Jobs that rely on interpersonal interaction will not be readily replaced by machines. However, to succeed at these jobs, sociobehavioral skills—acquired in one's early years and shaped throughout one's lifetime—must be strong. Human capital is important because there is now a higher premium on adaptability.

Solutions are available. For example, to prepare for the changing nature of work countries must boost their investment in early childhood development. This is one of the most effective ways to build valuable skills for future labor markets. Countries can also boost human capital by ensuring that schooling results in learning. Important adjustments in skills to meet the demands of the changing nature of work are also likely outside compulsory schooling and formal jobs. Countries can, for example, utilize tertiary education and adult learning more effectively.

One reason governments do not invest in human capital is the lack of political incentives. Few data are publicly available on whether health and education systems are generating human capital. This gap hinders the design of effective solutions, the pursuit of improvement, and the ability of citizens to hold their governments accountable. The World Bank's human capital project, described in this study, is designed to address the shortcomings of political incentives and provide the impetus for investing in human capital.

Social assistance and insurance systems should also be adapted to the changing nature of work. The concept of progressive universalism could be a guiding principle in covering more people, especially in the informal economy. When social protection is established, flexible labor regulation eases work transitions.

The current social contract is broken in most emerging economies, and it is looking increasingly out of date in some advanced economies as well. A new social contract should include investing in human capital to generate more opportunities for workers to find better jobs. This will improve the job prospects for newborns or schoolchildren.

How will governments raise the additional resources needed to invest in human capital and advance social inclusion? The share of tax revenue in low-income countries is half that of high-income countries (figure O.6). Investments in human capital, basic social protection (including community health workers in some developing countries), and productive opportunities for youth are likely to have fiscal costs of 6–8 percent of gross domestic product (GDP). This is an ambitious goal. Increasing tax revenue, however, should go hand in hand with improving public service delivery. If not, increasing tax rates will only spur further public discontent.

Most of the required fiscal resources are likely to come from improved capacity in tax administration and policy changes, particularly to value added taxes and through expansion of the tax base. Sub-Saharan African countries could raise, on average, from 3 to 5 percent of GDP in additional revenues through reforms that improve the efficiency of the current tax systems.[5] Closing tax exemptions and converging toward a uniform tax rate in value added tax could raise further revenues. In Costa Rica and Uruguay, such revenues could amount to more than 3 percent of GDP.

Other taxes and savings could also contribute to the financing of human capital. Saudi Arabia adopted excise taxes in 2017: 50 percent on soft drinks and 100 percent on energy drinks, tobacco, and tobacco products.

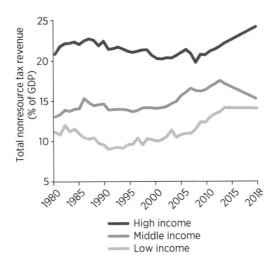

FIGURE O.6 Tax revenues are lower in developing countries

Sources: WDR 2019 team, based on UNU-WIDER's Government Revenue Dataset, 2017; World Bank data.

Note: GDP = gross domestic product.

It is estimated that nationally efficient carbon pricing policies would raise more than 6 percent of GDP in China, the Islamic Republic of Iran, Russia, and Saudi Arabia.[6] Taxes on immovable property could raise an additional 3 percent of GDP in middle-income countries and 1 percent in poor countries.[7]

Age-old tax avoidance and evasion schemes by firms and individuals need to be tackled as well. Four out of five Fortune 500 companies operate one or more subsidiaries in countries broadly perceived to operate preferential corporate tax regimes—often referred to as "tax havens." As a result, estimates suggest that governments worldwide may miss out on US$100–$240 billion in annual revenue, which is equivalent to 4–10 percent of the global corporate income tax revenue. The increasingly digital nature of business only creates more opportunities for tax avoidance. Generating revenue from new kinds of assets, such as user data, makes it increasingly unclear how or where value is created for tax purposes.

Organization of this study

The first chapter of this study looks at the impact of technology on jobs. In some sectors, robots are replacing workers. In other sectors, robots are enhancing worker productivity. And in further sectors, technology is creating jobs as it shapes the demand for new goods and services. These disparate effects of technology render the economic predictions of technology-induced job losses basically useless. Predictions sensationalize the impact of technology and stir fears, especially among middle-skill workers in routine jobs.

Technology does, however, change the demand for skills. Since 2001, the share of employment in occupations heavy in nonroutine cognitive and sociobehavioral skills has increased from 19 to 23 percent in emerging economies and from 33 to 41 percent in advanced economies. The payoffs to these skills, as well as to combinations of different skill types, are also increasing in those economies. But the pace of innovation will determine whether new sectors or tasks emerge to counterbalance the decline of old sectors and routine jobs as technology costs decline. Meanwhile, whether the cost of labor remains low in emerging economies in relation to capital will determine whether firms choose to automate production or move elsewhere. Chapter 1 sets out a model for the changing nature of work.

One feature of the current wave of technological progress is that it has made the boundaries of firms more permeable and has accelerated the emergence of superstar firms. Such firms have a beneficial effect on the demand for labor by boosting production and employment. These firms are also large integrators of young, innovative firms, often benefiting small businesses by connecting them with larger markets. But large firms, particularly firms in the digital economy, also pose risks. Regulations often fail to address the challenges created by new types of businesses in the digital economy. Antitrust frameworks also have to adjust to the impact of network effects on competition. Tax systems in many ways no longer fit their purposes as well. Chapter 2 examines how technological change affects the nature of the firm.

At the economywide level, human capital is positively correlated with the overall level of adoption of advanced technologies. Firms with a higher share of educated workers do better at innovating. Individuals with stronger human capital reap higher economic returns from new technologies. By contrast, when technological disruptions are met with inadequate human capital, the existing social order may be undermined. Chapter 3 addresses the link between human capital accumulation and the future of work, looking more closely at why governments need to invest in human capital and why they often fail to do so.

Chapter 3 also introduces the World Bank's new human capital project. To ensure effective policy design and delivery, more information and better measurement of foundational human capital are needed, even when there is full willingness to invest in human capital. The project has three components: a global benchmark—the human capital index; a program of measurement and research to inform policy action; and a program of support for country strategies to accelerate investment in human capital.

The index is measured in terms of the amount of human capital that a child born in 2018 can expect to attain by the end of secondary school, taking into account the risks of poor health and poor education that prevail in the country in which the child was born during that same year. In other words, it measures the productivity of the next generation of workers relative to a benchmark of complete education and full health. For example, in many education systems a year of schooling produces only a fraction of the learning that is possible (figure O.7). Chapter 3 presents cross-country comparisons for 157 economies globally.

Part of the ongoing skills readjustment is happening outside of compulsory education and formal jobs. But where? Chapter 4 answers this question by exploring three domains—early childhood, tertiary education, and adult learning outside jobs—where people acquire specific skills that the changing nature of work requires.

Investments in early childhood, including in nutrition, health, protection, and education, lay strong foundations for the future acquisition of higher-order cognitive and sociobehavioral skills. From the prenatal period to age 5, the brain's ability to learn from experience is at its highest. Individuals who acquire such skills in early childhood are more resilient to uncertainty later in life. Tertiary education is another opportunity for individuals to acquire the higher-order general cognitive skills—such as complex problem-solving, critical thinking, and advanced communication—that are so important to the changing nature of work but cannot be acquired through schooling alone.

As for the current stock of workers, especially those who cannot go back to school or to university, reskilling and upskilling those who are not in school or in formal jobs must be part of the response to technology-induced labor market disruption. But only rarely do adult learning programs get it right. Adults face various binding constraints that limit the effectiveness of traditional approaches to learning. Better diagnosis and evaluation of adult learning programs, along with better design and better delivery of those programs, are needed. Chapter 4 explores these issues in greater detail.

Work is the next venue for human capital accumulation after school. Chapter 5 evaluates how successful economies have been in generating human capital at work. Advanced economies have higher returns to work than emerging economies. A worker in an emerging economy is more likely than a worker in an advanced economy to find herself in a manual occupation that is composed largely of physical tasks. An additional year of work in cognitive professions increases wages by 3 percent, whereas for manual occupations the figure is 2 percent. Work provides a venue for a prolonged acquisition of skills after school—but such opportunities are relatively rare in emerging economies.

FIGURE O.7 Learning varies across emerging economies

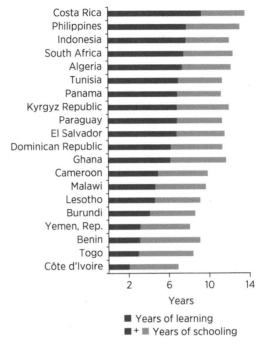

Sources: WDR 2019 team, based on Kim (2018); Filmer et al. (2018).

Governments can raise the returns to work by creating formal jobs for the poor. They can do this by nurturing an enabling environment for business, investing in entrepreneurship training for adults, and increasing access to technology. The payoff to women's participation in the workforce is significantly lower than for men—in other words, women acquire significantly less human capital than men do from work. To bridge that gap, governments could seek to remove limitations on the type or nature of work available to women and eliminate rules that would limit women's property rights. Workers in rural areas face similar challenges when it comes to accumulating human capital after school. There is some scope for improving the returns to work by reallocating labor from villages to cities. However, in rural areas technology can be harnessed to increase payoffs to work by increasing agricultural productivity.

Uncertain labor markets call for strengthening social protection. This topic is explored in chapter 6. Traditional provisions of social protection based on steady wage employment, clear definitions of employers and employees, and a fixed point of retirement are becoming increasingly obsolete. In developing countries, where informality is the norm, this model has been largely aspirational.

Spending on social assistance should be complemented with insurance that does not fully depend on having formal wage employment. The aim of this approach is to expand coverage while giving priority to the poorest people. As people become better protected through enhanced social assistance and insurance, labor regulation could, where appropriate, be rebalanced to facilitate work transitions.

Changes in the nature of work, compounded by rising aspirations, make it essential to increase social inclusion. To do so, a social contract should have at its center equality of opportunity. Chapter 7 considers potential elements of a social contract, which include investing early in human capital, taxing firms, expanding social protection, and increasing productive opportunities for youth.

To achieve social inclusion, some emerging economy governments will have to devise ways to increase revenue. Chapter 7 describes how governments can create fiscal space through a mix of additional revenues from existing and new funding sources. Potential sources of revenue are imposing value added taxes, excise taxes, and carbon taxes; charging platform companies taxes equal to what other companies are paying; and revisiting energy subsidies.

* * *

Simeon Djankov and Federica Saliola led the WDR 2019 team. The core team is composed of Ciro Avitabile, Rong Chen, Davida Connon, Ana Paula Cusolito, Roberta Gatti, Ugo Gentilini, Asif Mohammed Islam, Aart Kraay, Shwetlena Sabarwal, Indhira Vanessa Santos, David Sharrock, Consuelo Jurado Tan, and Yucheng Zheng. Paul Romer, former Chief Economist; Michal Rutkowski, Senior Director of the Social Protection and Jobs Global Practice; and Shantayanan Devarajan, Acting Chief Economist, provided guidance.

Notes

1. Marx (1867).
2. Keynes ([1930] 1963).
3. Brynjolfsson et al. (2008).
4. Clausing (2016).
5. IMF (2017).
6. Parry, Veung, and Heine (2014).
7. Norregaard (2013).

References

Brynjolfsson, Erik, Andrew McAfee, Michael Sorell, and Feng Zhu. 2008. "Scale without Mass: Business Process Replication and Industry Dynamics." Harvard Business School Technology and Operations Management Unit Research Paper No. 07-016, Cambridge, MA.

Clausing, Kimberly A. 2016. "The Effect of Profit Shifting on the Corporate Tax Base in the United States and Beyond." *National Tax Journal* 69 (4): 905–34.

Djankov, Simeon, Rafael la Porta, Florencio Lopez-de-Silanes, and Andrei Shleifer. 2002. "The Regulation of Entry." *Quarterly Journal of Economics* 118 (1): 1–37.

Filmer, Deon, Halsey Rogers, Noam Angrist, and Shwetlena Sabarwal. 2018. "Learning-Adjusted Years of Schooling (LAYS): Defining a New Macro Measure of Education." Policy Research Working Paper 8591, World Bank, Washington, DC.

IMF (International Monetary Fund). 2017. "Tackling Inequality." *Fiscal Monitor*, World Economic and Financial Surveys, IMF, Washington, DC, October.

Keynes, John Maynard. [1930] 1963. "Economic Possibilities for Our Grandchildren." In *Essays in Persuasion*, 358–73. New York: W. W. Norton. http://www.econ.yale.edu/smith/econ116a/keynes1.pdf.

Kim, Jim Yong. 2018. "The Human Capital Gap: Getting Governments to Invest in People." *Foreign Affairs* (July/August). https://www.foreignaffairs.com/articles/2018-06-14/human-capital-gap.

Marx, Karl. 1867. *Das Kapital: Kritik der politischen Ökonomie*. Hamburg: Verlag von Otto Meissner.

Norregaard, John. 2013. "Taxing Immovable Property: Revenue Potential and Implementation Challenges." IMF Working Paper WP/13/129, International Monetary Fund, Washington, DC, May 29.

Parry, Ian W. H., Chandara Veung, and Dirk Heine. 2014. "How Much Carbon Pricing Is in Countries' Own Interests? The Critical Role of Co-benefits." IMF Working Paper WP/14/174, International Monetary Fund, Washington, DC, September 17.

CHAPTER 1

The changing nature of work

From their beginning, robots were intended to replace humans in the workplace. In fact, Karel Čapek, the Czech writer who invented the word *robot* in 1920, used the Slavic-language word for work, *robota*, to make it clear what these machines would be used for. Over the last century, machines have replaced workers in many tasks. On balance, however, technology has created more jobs than it has displaced. Technology has brought higher labor productivity to many sectors by reducing the demand for workers for routine tasks. And yet in doing so, it has opened doors to new sectors once imagined only in the world of science fiction.

As technology advances, firms adopt new methods of production, markets expand, and societies evolve. Firms rely on new technologies to better use capital, overcome information barriers, outsource, and innovate. New technologies allow for more efficient management of the operations of firms: firms hire workers in one location to produce parts, in another location to assemble, and in a third location to sell. Meanwhile, consumers enjoy a wider range of products at lower prices.

In today's economy, market opportunities are growing for all participants. Some platform firms[1] are creating new marketplaces to trade goods or services. Even small firms are global. And they are growing faster. The firms selling on eBay in Chile, Jordan, Peru, and South Africa are younger than the firms in offline markets.[2] In China, start-ups are dominant on the Alibaba platform.[3] Societies benefit as technology increases the options for service delivery and for citizens to hold their governments accountable.

Workers, firms, and governments are building new comparative advantages as conditions change. For example, by being the first to adopt 3-D technologies, Danish firms strengthened their hold on the global market for hearing aid products in the 2000s.[4] The Indian government invested in technical universities across the country, and subsequently India became a world leader in high-tech sectors. By integrating into global value chains, Vietnamese workers developed their foreign-language abilities, building additional human capital that allows them to expand into other markets.

Notwithstanding the opportunities, however, there are still disruptions. The declining cost of machines especially puts at risk those workers in low-skill jobs engaged in routine tasks. These are the occupations most susceptible to automation. Displaced workers are likely to compete with (other) low-skill workers for jobs with low wages. Even when new jobs are created, retooling is costly, and often impossible.

The resulting displacement of workers generates anxiety, just as in the past. In 1589, Queen Elizabeth I of England was alarmed when clergyman William Lee applied for a royal patent for a knitting machine: "Consider thou what the invention would do to my poor subjects," she pointed out. "It would assuredly bring them to ruin by depriving them of employment."[5] In the 1880s, the Qing dynasty fiercely opposed constructing railways in China, arguing that the loss of luggage-carrying jobs might lead to social turmoil.[6] Earlier in the 19th century, the Luddites sabotaged machines to defend their jobs in England, despite the overall economic growth fueled by steam power.

Fears of robot-induced unemployment have dominated discussions about the future of work. Nowhere are these fears more evident than in the industrial sector. The decline in industrial employment in some high-income economies over the last two decades is an established trend. The Republic of Korea, Singapore, Spain, and the United Kingdom are among the countries in which the share of industrial employment dropped more than 10 percentage points. But this trend mainly reflects a shift in employment from manufacturing to services as those countries grow. By contrast, millions of industrial jobs have been created in developing countries since the late 1980s. Indeed, the share of industrial employment has increased significantly in a few emerging markets such as Cambodia and Vietnam. On average, the share of industrial employment has remained stable in developing countries, despite the many predictions of job losses resulting from technology.

That said, technology is disrupting the demand for skills. Globally, private returns to education—about 9 percent a year—remain high despite the significant expansion in the supply of skilled labor. Returns to tertiary education are almost 15 percent a year. Individuals with more advanced skills are taking better advantage of new technologies to adapt to the changing nature of work. For example, returns to primary schooling in India increased during the Green Revolution of the 1960s and 1970s, with more educated farmers adopting new technologies.

Technology has the potential to improve living standards, but its effects are not manifesting themselves equally across the globe. The process of job creation works societywide—and not just for the few—only when the rules of the game are fair. Workers in some sectors benefit handsomely from technological progress, whereas those in others are displaced and have to retool to survive. Platform technologies create huge wealth but place it in the hands of only a few people.

Irrespective of technological progress, persistent informality continues to pose the greatest challenge for emerging economies. Informal employment remains at more than 70 percent in Sub-Saharan Africa and 60 percent in South Asia and at more than 50 percent in Latin America. In India, the informal sector has remained at around 90 percent, notwithstanding fast economic growth and technology adoption. Both wages and productivity are significantly lower in the informal sector. Informal workers have neither health insurance nor social protection. Technology may prevent Africa and South Asia from industrializing in a manner that moves workers to the formal sector.

Progress in the context of the formal–informal worker divide must be reevaluated because of the changing nature of work. Economic growth depends on human capital accumulation and infrastructure that responds to the needs of education, health, and business. Enhanced social protection that applies no matter the form of labor contract is also ripe for consideration.

Technology generates jobs

"They're always polite, they always upsell, they never take a vacation, they never show up late, there's never a slip-and-fall, or an age, sex, or race discrimination case," said Andrew Puzder, then chief executive of Hardee's Food Systems Inc., a restaurant chain headquartered in Tennessee. He was talking about swapping employees for machines.[7] Statements like these give workers reason to worry.

The advent of a jobless economy raises concern because tasks traditionally performed by humans are being—or are at risk of being—taken over by robots, especially those enabled with artificial intelligence. The number of robots operating worldwide is rising quickly. By 2019, 1.4 million new industrial robots will be in operation, raising the total to 2.6 million worldwide.[8] Robot density per worker in 2018 is the highest in Germany, Korea, and Singapore. Yet in all of these countries, despite the high prevalence of robots, the employment rate remains high.

Young workers may be more affected by automation than older workers. Although the adoption of robots did not have any substantial net effect on employment in Germany, it reduced the hiring of young entrants.[9] For this reason, the effects of automation can be different in countries that are aging compared with those that have young populations and anticipate large numbers of new labor market entrants.

Yes, robots are replacing workers, but it is far from clear to what extent. Overall, technological change that replaces routine work is estimated to have created more than 23 million jobs across Europe from 1999 to 2016, or almost half of the total increase in employment over the same period. Recent evidence for European countries suggests that although technology may be replacing workers in some jobs, overall it raises the demand for labor.[10] For example, instead of hiring traditional loan officers, JD Finance, a leading fintech platform in China, created more than 3,000 risk management or data analysis jobs to sharpen algorithms for digitized lending.

Technological progress leads to the direct creation of jobs in the technology sector. People are increasingly using smartphones, tablets, and other portable electronic devices to work, organize their finances, secure and heat their homes, and have fun. Workers create the online interfaces that drive this growth. With consumer interests changing fast, there are more opportunities for people to pursue careers in mobile app development and virtual reality design.

Technology has also facilitated the creation of jobs through working online or joining the so-called gig economy. Andela, a U.S. company that specializes in training software developers, has built its business model on the digitization of Africa. It has trained 20,000 software programmers across Africa using free online learning tools. Once qualified, programmers work with Andela directly or join other Andela clients across the world. The company aims to train 100,000 African software developers by 2024. Ninety percent of its workers are in Lagos, Nigeria, with other sites in Nairobi, Kenya, as well as Kampala, Uganda.

Technology increases proximity to markets, facilitating the creation of new, efficient value chains. In Ghana, Farmerline is an online platform that communicates with a network of more than 200,000 farmers in their native languages via mobile phone. It provides information on the weather and market prices, while collecting data for buyers, governments, and development partners. The company is expanding to include credit services.

During this process of technology adoption some workers will be replaced by technology. Workers involved in routine tasks that are "codifiable" are the most vulnerable. The examples are numerous. More than two-thirds of robots are employed in the automotive, electrical/electronics, and metal and machinery industries. Based in China, Foxconn Technology Group, the world's largest electronics assembler, cut its workforce by 30 percent when it introduced robots into the production process. When robots are cheaper than the existing manufacturing processes, firms become more amenable to relocating production closer to consumer markets. In 2017 3-D printing technologies enabled the German company Adidas to establish two "speed factories" for shoe production: one in Ansbach, Germany, and the other in Atlanta in the United States, eliminating more than 1,000 jobs in Vietnam. In 2012 the Dutch multinational technology company Philips Electronics shifted production from China back to the Netherlands.

Some service jobs are also vulnerable to automation. Mobileye of Israel is developing driverless vehicle navigation units. Baidu, the Chinese technology giant, is working with King Long Motor Group, China, to introduce autonomous buses in industrial parks. Financial analysts, who spend much of their time conducting formula-based research, are also experiencing job cuts: Sberbank, the largest bank in the Russian Federation, relies on artificial intelligence to make 35 percent of its loan decisions, and it anticipates raising that rate to 70 percent in less than five years.[11] "Robot lawyers" have already replaced 3,000 human employees in Sberbank's legal department. The number of back-office employees will shrink to 1,000 by 2021, down from 59,000 in 2011. Ant Financial, a fintech firm in China, uses big data to assess loan agreements instead of hiring thousands of loan officers or lawyers.

Nevertheless, it is impossible to put a figure on the level of job displacement that will take place overall. Even the most well-known economists have experienced little success with this exercise. In 1930 John Maynard Keynes declared that technology would usher in an age of leisure and abundance within a hundred years. He mused that everyone would have to do some work if they were to be content, but that three hours a day would be quite enough.[12] The world in 2018 is far from this kind of reality.

Although quantifying the impact of technological progress on job losses continues to challenge economists, estimates abound. Those estimates vary widely (figure 1.1). For Bolivia, job automation estimates range from 2 to 41 percent. In other words, anywhere from 100,000 to 2 million Bolivian jobs may be automated in 2018. The range is even wider for advanced economies. In Lithuania, from 5 to 56 percent of jobs are at risk of being automated. In Japan, from 6 to 55 percent of jobs are thought to be at risk.

FIGURE 1.1 Estimates of the percentage of jobs at risk from automation vary widely

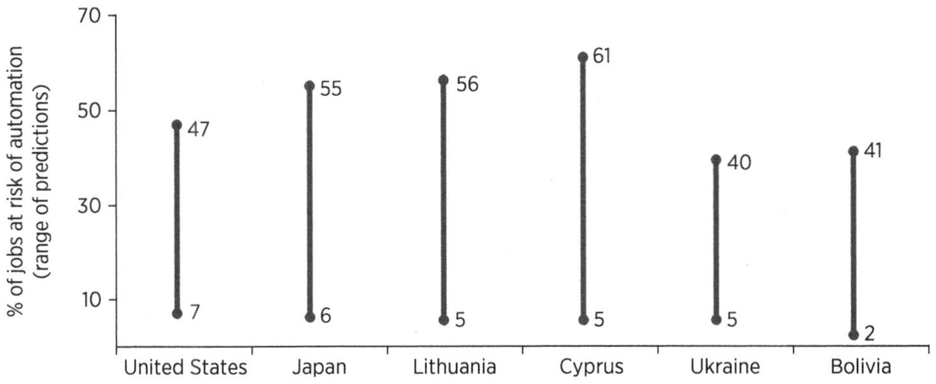

Sources: WDR 2019 team, based on World Bank (2016); Arntz, Gregory, and Zierahn (2016); David (2017); Hallward-Driemeier and Nayyar (2018).

Note: The figures represent the highest and lowest estimates of the percentage of jobs at risk of automation in economies for which more than one estimate has been produced by different studies. A job is at risk if its probability of being automated is greater than 0.7.

The wide range of predictions illustrates the difficulty of estimating technology's impact on jobs. Most estimates rely on automation probabilities developed by machine learning experts at the University of Oxford. The experts were asked to categorize a sample of 70 occupations taken from the O*NET online job database used by the U.S. Department of Labor as either strictly automatable or not (1–0).[13] Relying on these probabilities, initial estimates placed 47 percent of U.S. occupations at risk of automation. Basing probabilities on the opinion of experts is instructive but not definitive. Moreover, using one country's occupational categories to estimate possible job losses from automation elsewhere is problematic.

Job loss predictions do not accurately incorporate technology absorption rates, which are often painstakingly slow and differ not only between countries but also across firms within countries. The absorption rate therefore affects the potential for technology to destroy jobs. The use of mobile telephony, for example, spread faster than earlier technologies, but the Internet has been comparatively slow to take hold in many cases, particularly among firms in the informal sector. The uptake of mechanization in agriculture presents a similar picture. Persistent trade barriers, the relatively low cost of labor compared with that of agricultural machinery, and poor information all contribute to the low rates of mechanization in low-income and some middle-income countries. Even for the textile industry's spinning jenny, the relatively low cost of labor delayed its introduction in France and India—in 1790 France had only 900 spinning jennies compared with 20,000 in Great Britain.[14] The prevalence of automation versus labor continues to vary across and within countries, depending on the context.

How work is changing

It is easier to assess how technology shapes the demand for skills and changes production processes than it is to estimate its effect on job losses. Technology is changing the skills being rewarded in the labor market. The premium is rising for skills that cannot be replaced by robots—general cognitive skills such as critical thinking and sociobehavioral skills such as managing and recognizing emotions that enhance teamwork. Workers with these skills are more adaptable in labor markets. Technology is also disrupting production processes by challenging the traditional boundaries of firms, expanding global value chains, and changing the geography of jobs. Finally, technology is changing how people work, giving rise to the gig economy in which organizations contract with independent workers for short-term engagements.

Technology is disrupting the demand for three types of skills in the workplace. First, the demand for nonroutine cognitive and sociobehavioral skills appears to be rising in both advanced and emerging economies. Second, the demand for routine job-specific skills is declining. And, third, payoffs to combinations of different skill types appear to be increasing. These changes show up not just through new jobs replacing old jobs, but also through the changing skills profile of existing jobs (figure 1.2).

Since 2001, the share of employment in occupations intensive in nonroutine cognitive and sociobehavioral skills has increased from 19 to 23 percent

FIGURE 1.2 Sociobehavioral skills are becoming more important

Job requirements of a Hilton Hotel management trainee in Shanghai, China

1986

Shanghai Hilton International
上海静安希尔顿酒店

上海静安希尔顿酒店招聘启事

本五星级酒店是国际希尔顿公司在中国管理的第一个企业，属全独资外资合作经营。楼高 43 层，客房 800 间，中外餐厅酒吧 8 个，设备极其豪华，位于上海静安区。将在明年年中以后开始营业，届时将成为国际希尔顿公司在世界 50 多个国家，超过 100 个酒店的大家庭中的一员。现在招聘受训管理人员，条件如下：

一、 素质：品质优秀，态度良好，勤奋好学。
二、 年龄：20 岁至 26 岁（1959 年-1966 年出生）。
三、 文化程度：大学或大专毕业。
四、 外语程度：英语"新概念"第二册以上，会话流利；同时能操其他外语者，优先考虑。
五、 健康状况：优良。
六、 其他：住处最好在静安区附近。

如具备上述条件，并有意尝试在我酒店取得发展者，请在八月二十、二十一、二十二日三天上午 9：00-11：00，下午 2：00-5：00，带学历证明，本人近期照片一张和五元报名费，到茂名北路 40 号新群中学报名，如为在职者，应持所属单位许可证明。

- Excellent character, willingness to learn
- Ages 20–26
- Bachelor's degree or associate degree
- Proficient in English
- Good health
- Live close to the hotel location

2018

Management Trainee

Front Office serving Hilton brands are always working on behalf of our Guests and working with other Team Members. To successfully fill this role, you should maintain the attitude, behaviors, skills, and values that follow:

- Previous experience in a customer-focused industry
- Positive attitude and good communication skills
- Commitment to delivering a high level of customer service
- Excellent grooming standards
- Ability to work on your own and as part of a team
- Competent level of IT proficiency

- Positive attitude and good communication skills
- Ability to work independently and as part of a team
- Competent level of IT proficiency
- Four-year university degree with at least two years of experience

Sources: 1986: *Wenhui News*, August 17, 1986, http://www.sohu.com/a/194532378_99909679; 2018: https://www.hosco.com/en/job/waldorf-astoria-shanghai-on-the-bund/management-trainee-front-office.

Note: IT = information technology.

in emerging economies and from 33 to 41 percent in advanced economies. In Vietnam, within a given industry workers performing nonroutine analytical tasks earn 23 percent more than those performing tasks that are nonanalytical, noninteractive, and nonmanual; those undertaking interpersonal tasks earn 13 percent more.[15] In Armenia and Georgia, the earnings premium for problem solving and learning new skills at work is close to 20 percent.[16]

Robots may complement workers who engage in nonroutine tasks that require advanced analytical, interpersonal, or manual skills requiring significant dexterity—for instance, teamwork, relationship management, people management, and caregiving. In these activities, people must interact with one another on the basis of tacit knowledge. Designing, producing art, conducting research, managing teams, nursing, and cleaning have proven to be hard tasks to automate. Robots have for the most part struggled to replicate these skills to compete with workers.

Machines replace workers most easily when it comes to routine tasks that are codifiable. Some of these tasks are cognitive, such as processing payrolls or bookkeeping. Others are manual or physical, such as operating welding machines, assembling goods, or driving forklifts. These tasks are easily automated. In Norway, the adoption by firms of information and communications technologies benefited skilled workers in executing nonroutine abstract tasks but replaced unskilled workers.[17]

Payoffs for combinations of different skill types are also increasing. The changing nature of work demands skill sets that improve the adaptability of workers, allowing them to transfer easily from one job to another. Across countries, both higher-order cognitive (technical) skills and sociobehavioral skills are consistently ranked among the skills most valued by employers. Employers in Benin, Liberia, Malawi, and Zambia rank teamwork, communication, and problem-solving skills as the most important set of skills after technical skills.[18]

Even within a given occupation, the impact of technology on the skills required to perform a job is changing—but not always in the direction one might expect. In Chile, the adoption of sophisticated computer software for client management and business operations between 2007 and 2013 decreased the demand for workers to complete abstract tasks and increased the demand for workers to complete routine manual tasks. As a result, there was a reallocation of employment from skilled workers to administrative, unskilled production workers.[19]

In advanced economies, employment has been growing fastest in high-skill cognitive occupations and low-skill occupations that require dexterity. By contrast, employment has shifted away from middle-skill occupations such as machine operators. This is one of the factors that may translate into rising inequality in advanced economies. Both middle- and low-skill workers could see falling wages—the former because of automation; the latter because of increased competition.

Few studies have been made of emerging economies, but some of those that have been made reveal similar changes in employment. In middle-income

European countries such as Bulgaria and Romania, the demand for workers in occupations involving nonroutine cognitive and interpersonal skills is rising, while the demand for workers in lower-skill nonroutine manual occupations has remained steady.[20] The use of routine cognitive skills has also increased in Botswana, Ethiopia, Mongolia, the Philippines, and Vietnam.[21] Studies observe that the demand for nonroutine cognitive and interpersonal skills is largely rising much faster than for other skills. High-skill workers are gaining with technological change, whereas low-skill workers—especially those in manual jobs—seem to be losing out.

Other studies show that changes in employment have been positive. In Argentina, the adoption of information and communications technologies in manufacturing increased employment turnover: workers were replaced, occupations were eliminated, new occupations were created, and the share of unskilled workers fell. However, employment levels increased across all skill categories.[22]

Technology is also disrupting production processes, challenging the traditional boundaries of firms and expanding global value chains. In doing so, technology changes the geography of jobs. Other waves of technological change have done the same. The Industrial Revolution, which mechanized agricultural production, automated manufacturing, and expanded exports, led to the mass migration of labor from farms to cities. The advent of commercial passenger planes expanded tourism from local holiday destinations in Northern Europe to new foreign resorts on the Mediterranean Sea. Thousands of new jobs were created in new locations.

Improvements in transcontinental communications technologies, along with the fall in transportation costs, have expanded global value chains toward East Asia. But many other factors beyond technology also matter for outsourcing. The Philippines overtook India in 2017 in terms of market share in the call center business at least in part because of the country's lower taxes.

Meanwhile, technology is enabling clusters of business to form in underdeveloped rural areas. In China, rural micro e-tailers began to emerge in 2009 on Taobao.com Marketplace. Owned by Alibaba, it is one of the largest online retail platforms in China. These clusters—"Taobao Villages"—spread fast, from just 3 in 2009 to 2,118 across 28 provinces in 2017. In 2017 490,000 shops were online. Although sales have been strongest in traditional goods such as apparel, furniture, shoes, luggage, leather goods, and auto accessories, sellers are diversifying their offerings to include high-tech goods such as drones.

Online work platforms are eliminating many of the geographical barriers previously associated with certain tasks. Bangladesh contributes 15 percent to the global labor pool online by means of its 650,000 freelance workers.[23] Indiez, founded in 2016 in India, takes a team-based approach to online freelancing. The platform provides a remotely distributed community of talent—mainly from India, Southeast Asia, and Eastern Europe—that works together on tech projects for clients anywhere in the world. Clients include the pizza restaurant chain Domino's India, as well as the Indian multinational conglomerate Aditya Birla Group. Wonderlabs in Indonesia follows a similar model.

Finally, technology is changing how people work and the terms under which they work. Instead of the once standard long-term contracts, digital technologies are giving rise to more short-term work, often via online work platforms. These so-called gigs make certain kinds of work more accessible on a more flexible basis. More widespread access to digital infrastructure—via laptops, tablets, and smartphones—provides an enabling environment in which on-demand services can thrive. Examples range from grocery delivery and driving services to sophisticated tasks such as accounting, editing, and music production. Asuqu in Nigeria connects creatives and other experts with businesses across Africa. Crew Pencil works in the South African movie industry. Tutorama, based in the Arab Republic of Egypt, connects students with local private tutors. In Russia, students work as Yandex drivers whenever they can fit it in to their university schedules. They identify peak hours in different locations to achieve the highest level of passenger turnover.

It is difficult to estimate the size of the gig economy. Where data exist, the numbers are still small. Data from Germany and the Netherlands indicate that only 0.4 percent of the labor force of those countries is active in the gig economy. Worldwide, the total freelancer population is estimated at around 84 million, or less than 3 percent of the global labor force of 3.5 billion.[24] A person counted as a freelancer may also engage in traditional employment. In the United States, for example, more than two-thirds of the 57.3 million freelancers also hold a traditional job, using freelancing to supplement their income.[25] The best estimate is that less than 0.5 percent of the active labor force participates in the gig economy globally, with less than 0.3 percent in developing countries.

Changes in the nature of work are in some ways more noticeable in advanced economies where technology is widespread and labor markets start from higher levels of formalization. However, emerging economies have been grappling with many of the same changes for decades. As noted earlier, informality persists on a vast scale in emerging economies—as high as 90 percent in some low- and middle-income countries—notwithstanding technological progress. With some notable exceptions in Eastern Europe, informality has been hard to tackle. In countries such as El Salvador, Morocco, and Tanzania only one out of five workers is in the formal sector. On average, two out of three workers in emerging economies are informal workers (figure 1.3).

The prevalence of informality predates the new millennium wave of technological change. Various programs for reducing informality, inspired by Hernando de Soto's *The Other Path: The Economic Answer to Terrorism* (2002), have yielded limited progress. The reason is the onerous regulations, taxes, and social protection schemes that give businesses no incentive to grow.

Because recent technological developments are blurring the divide between formal and informal work, there is something of a convergence

FIGURE 1.3 Two out of three workers in emerging economies are in the informal economy (selected countries)

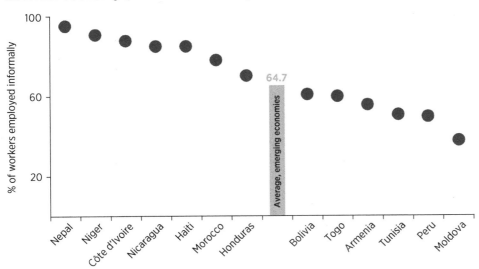

Source: WDR 2019 team, using household and labor force survey data from the World Bank's International Income Distribution Data Set.

Note: The figure shows selected countries with the highest rates of informal employment. A person is identified as an informal worker if he or she does not have an employment contract, social security, and health insurance, and is not a member of a labor union. The estimates are for the latest available year for each country, ranging from 2010 to 2016.

in the nature of work between advanced and emerging economies. Labor markets are becoming more fluid in advanced economies, while informality is persisting in emerging economies. Most of the challenges faced by short-term or temporary workers, even in advanced economies, are the same as those faced by workers in the informal sector. Self-employment, informal wage work with no written contracts or protections, and low-productivity jobs more generally are the norm in most of the developing world. These workers operate in a regulatory gray area, with most labor laws unclear on the roles and responsibilities of the employer versus the employee. This group of workers often lacks access to benefits. There are no pensions, no health or unemployment insurance schemes, and none of the protections provided to formal workers.

This type of convergence is not what was expected in the 21st century. Traditionally, economic development has been synonymous with formalization. This is reflected in the design of social protection systems and labor regulations. A formal wage employment contract is still the most common basis for the protections afforded by social insurance programs and by regulations such as those specifying a minimum wage or severance pay. Changes in the nature of work caused by technology shift the pattern of demanding workers' benefits from employers to directly demanding welfare benefits from the state. These changes raise questions about the ongoing relevance of current labor laws.

A simple model of changing work

Will robots turn the old Luddite fears of machines replacing workers into reality? Will massive automation mean that the old path of prosperity-through-industrialization, once taken by China, Japan, and the United Kingdom, is closing? How can public policy ensure that the evolution of work produces a world that is both more prosperous and more equitable?[26]

High labor costs in relation to capital—beyond a certain level—push firms to automate production or to move jobs to lower-cost countries (figure 1.4). This reduction in costs is achieved explicitly within a firm or implicitly through competition within a market. The relative cost of labor, not income, is emphasized because countries may have labor costs that do not align with their income level. This is the case, for example, in countries where low levels of human capital render workers unproductive, reducing exporting potential, or in countries where regulations significantly raise labor costs for formal employers.

A response to globalization is a greater shift in jobs to developing country cities, thereby reducing the overall relative costs of labor (and shifting the curve in figure 1.4 leftward). Automation leads to less demand for manufacturing workers everywhere (shifting the curve downward). Automation also changes the overall relationship between industrial employment and labor costs because it occurs more quickly in locations with high labor costs, assuming the incentive to reduce labor costs trumps other differences between locations (changing the shape of the curves in figure 1.4 from left-skewed to right-skewed).

Keynes understood that employment in the traditional sectors, especially agriculture, would decline enormously in the 20th century, but he failed to anticipate the explosion of new products that 21st-century workers would produce and consume. Most important of all, he failed to foresee the vast service economy that would employ workers in most wealthy countries. Digital technologies are enabling firms to automate, replacing labor with machines in production, and to innovate, expanding the number of tasks and products. The future of work will be determined by the battle between automation and innovation (figure 1.5). In response to automation, employment in old sectors declines. In response to innovation, new sectors or tasks emerge. The overall future of employment depends on both. It also depends on the labor and

FIGURE 1.4 Automation and globalization affect industrial employment

Source: Glaeser 2018.

Note: The curves are inverse U-shaped to reflect the empirical regularity that manufacturing employment constitutes a larger share of employment in middle-income countries; higher-income countries tend to specialize in services; and low-income countries have a relatively higher share of employment in agriculture.

skills intensity of the new sectors or tasks that emerge. These forces in turn affect wages.

For most of the last 40 years, human capital has served as a shield against automation, in part because machines are less adept at replicating more complex tasks. Low-skill and middle-skill workers have benefited less from technological change either because of higher susceptibility to automation or because of lower complementarities with technology.[27]

What is the result? Automation has disproportionately reduced the demand for less skilled workers, and the innovation process has generally favored the more educated. A big question is whether workers displaced by automation will have the required skills for new jobs created by innovation. This study focuses on the importance of human capital for the workforce of the future. Yet it is worth remembering that many innovations, such as Henry Ford's assembly lines, increased the demand for less skilled workers, while others, such as quartz watches, disproportionately destroyed jobs for higher-skill workers.

Automation and innovation are largely the unexpected by-products of a single breakthrough, such as the advent of the Internet, or the result of more targeted investments by companies that are seeking to either reduce labor costs or increase profits in new markets. If public regulations limit innovation, employment is more likely to fall.

In the mid-20th century, automation in the form of dishwashers and washing machines revolutionized homemaking, enabling millions of women to work outside the household. Women often found jobs in the service economy, which grew by providing yet more products and services, from caffe lattes to financial planning, and enabling an even finer division of labor such as personal trainers and financial market traders. A major question for this century is whether more of these services will become tradable and whether service workers will locate in the same metropolitan area as their clients.

The battle between innovation and automation is raging not just in the U.S. and European rust belts. Even though low-wage countries may not invest in the development of labor-saving innovations, they import labor-saving ideas from advanced economies. In fact, the mechanization of agriculture in emerging economies represents the largest global shift in work. Cities in emerging countries must generate abundant new jobs to employ the farmers displaced by the industrialization of agriculture. The declining costs of transportation

FIGURE 1.5 In the future, the forces of automation and innovation will shape employment

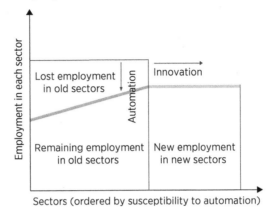

Source: Glaeser 2018.

Note: The ordering of the sectors in the figure should be understood as running from the most automatable to the least automatable, or from low-skill and middle-skill jobs to high-skill jobs where there is a decline in the relative demand for some less educated workers.

and connectivity (so-called globalization) enable these urban job markets to expand, as long as connectivity spreads more quickly than the automation of tradable goods production. So, although the growth of employment in emerging economies is supported by global value chains, automation may mean that African countries never experience mass industrialization.

The dramatic economic growth experienced by China, Japan, Korea, and Vietnam started with the fruits of globalization: manufacturing exports that competed effectively because of low labor costs. These countries chose to invest in infrastructure, special economic zones, and, above all, human capital, which generated a high-quality labor force connected to the outside world.

The transition of Shenzhen, China, from labor-intensive, low-cost manufacturing to high-skilled, technologically intense production illustrates the challenge that later industrializers are facing. They must compete not only with the high labor cost, capital-intensive producers of the wealthy West, but also with the moderate labor cost, technology-intensive producers of Asia and Eastern Europe. If robust global connections arrive too slowly in Africa, then industrialization may no longer be a plausible path to job creation. This threat strengthens the case for investing promptly in the precursors of globalization: education and transportation infrastructure.[28]

If African cities maintain the current model, employment will remain in the low-wage informal service sector. Changing the model depends significantly on investments in human capital (figure 1.6). In that case, Africa may urbanize as a services-producing economy, moving away from export earnings based on natural resources and agriculture.

Globalization increases the returns to human capital through higher labor productivity; some workers participate in export industries, and the shift of workers to those industries increases the demand for all kinds of labor (figure 1.6). This positive shift is meant to capture the positive experience of a poorer nation that has suddenly gained access to significant foreign direct investment. Of course, globalization may not always raise productivity across the board.

Likewise, the benefits of globalization will not accrue evenly. Globalization causes the variance in labor productivity to increase. Although productivity for subsistence farmers is low and relatively homogeneous, the returns to participating in a globalized economy are far more mixed. By investing strongly in raising the human capital of their citizens, governments increase their citizens' chances of success in global markets.

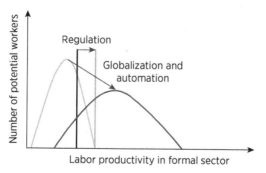

FIGURE 1.6 Human capital shapes productivity and wages in emerging economies

Source: Glaeser 2018.

Note: The vertical lines denote the minimum productivity level at which firms find it optimal to employ workers formally before the move toward globalization.

The vertical lines in figure 1.6 denote the minimum productivity level at which firms find it optimal to employ workers formally before the move toward globalization. A minimum wage, required benefits, and other taxes and regulations ensure that informality is appealing for all but the most productive workers before the economy grows. If regulations remained constant, globalization and automation would in many cases pull more workers into the formal sector by increasing their productivity. Yet this formal employment effect may be reduced if development prompts countries to impose more requirements on firms. Globalization raises incomes, but it may not do much to reduce informality if regulatory aspirations increase along with global connections. Indeed, informality could even rise if globalization sufficiently increases regulation.

Finally, policy makers have to think about risk management because of the predominance of informality in developing countries and the higher uncertainty associated with the changing nature of work. The large continuing presence of a vast informal service sector challenges risk management systems that function through employers. Financing pensions and other forms of insurance through payroll taxes levied on formal workers does little good if these workers represent only a small share of the workforce. Strong requirements also deter formalization.

This study emphasizes the importance for social inclusion for all workers regardless of how or where they work. Governments could try to strengthen social protection and reduce inequality through requirements or subsidies for employer-provided support such as a minimum wage, employer-provided health care, or protection against dismissal. Alternatively, governments could pursue the same goals through direct, state-provided support in the form of social assistance programs and subsidized universal social insurance or public jobs for, say, community health workers.

Both types of social policy promote equity. And both have costs. From the state's perspective, different combinations of regulations and public aid generate the same level of equity. Direct public aid generates implementation costs through waste and higher tax rates. Employer requirements deter hiring and could, if too stringent, raise inequity by increasing the share of workers who are either unemployed or in the informal sector.

Many developing countries initially chose to redistribute primarily through labor market regulations because the costs of distorting labor markets were low and the public capacity for social programs was limited. If automation pushes up the cost of distorting labor markets, and development improves the efficacy of the public sector, government should move away from regulation-based redistribution to direct social welfare support.

The future world of work is uncertain. Innovation may outpace automation. Globalization may move quickly enough that industrialization allows Africa to grow and prosper. Yet, given the considerable uncertainty about the future of employment, governments should rethink policies that deter job creation, and emphasize policies that protect the vulnerable while still encouraging employment.

Notes

1. Many Internet businesses or services use a platform or "two-sided market" model. The platforms match buyers with sellers or a service user with a provider. See World Bank (2016).
2. eBay Inc. (2013).
3. Chen and Xu (2015).
4. Freund, Mulabdic, and Ruta (2018).
5. McKinley (1958).
6. Zeng (1973).
7. Taylor (2016).
8. International Federation of Robotics, Frankfurt, https://ifr.org/.
9. Dauth et al. (2017).
10. Gregory, Salomons, and Zierahn (2016).
11. TASS (2017).
12. Keynes ([1930] 1963).
13. An algorithm was then used to extend that sample to categorize the remainder of the 632 U.S. occupational categories based on their task makeup. Where the probability of automation was greater than 0.7, that occupation was considered at risk (Frey and Osborne 2017).
14. Aspin (1964).
15. World Bank (2014).
16. World Bank (2015a, 2015b).
17. Akerman, Gaarder, and Mogstad (2015).
18. Arias, Santos, and Evans (2018).
19. Almeida, Fernandes, and Viollaz (2017).
20. Hardy, Keister, and Lewandowski (2018).
21. For East Asian countries, see Mason, Kehayova, and Yang (2018). For others, see World Bank (2016).
22. Brambrilla and Tortarolo (2018).
23. Aowsaf (2018).
24. This is a sum of various available statistics: 57.3 million, United States; 2 million, United Kingdom; 10 million, European Union; 15 million, India. These countries or regions are those in which freelancing is booming. The aggregated number likely represents a sizable portion of the global freelancer workforce.
25. Upwork (2017).
26. This section is based on Glaeser (2018).
27. Acemoglu and Autor (2011). In advanced economies, the replacement of labor with automation appears to be concentrated in middle-skill jobs, leading to the polarization of labor markets. This Report reveals that, at least so far, there is significant variation across developing countries in the relative employment growth of different occupations. In many countries, middle-skill jobs continue to grow in importance.
28. Education improves the ability of countries to take advantage of globalization. For example, successful exporters in developing countries tend to export higher-quality products, and high quality requires skills (Brambrilla, Lederman, and Porto 2012; Verhoogen 2008).

References

Acemoglu, Daron, and David H. Autor. 2011. "Skills, Tasks, and Technologies: Implications for Employment and Earnings." In *Handbook of Labor Economics*, Vol. 4, Part B, edited by Orley C. Ashenfelter and David Card, 1043–1171. San Diego, CA: North-Holland.

Akerman, Anders, Ingvil Gaarder, and Magne Mogstad. 2015. "The Skill Complementarity of Broadband Internet." *Quarterly Journal of Economics* 130 (4): 1781–1824.

Almeida, Rita K., Ana M. Fernandes, and Mariana Viollaz. 2017. "Does the Adoption of Complex Software Impact Employment Composition and the Skill Content of Occupations? Evidence from Chilean Firms." Policy Research Working Paper 8110, World Bank, Washington, DC.

Aowsaf, SM Abrar. 2018. "The Cost of Getting Paid." *Dhaka Tribune*, March 1. http://www.dhakatribune.com/business/2018/03/01/cost-getting-paid/.

Arias, Omar S., Indhira Santos, and David K. Evans. 2018. *The Skills Balancing Act in Sub-Saharan Africa: Investing in Skills for Productivity, Inclusion, and Adaptability.* Washington, DC: World Bank.

Arntz, Melanie, Terry Gregory, and Ulrich Zierahn. 2016. "The Risk of Automation for Jobs in OECD Countries: A Comparative Analysis." *OECD Social, Employment and Migration Working Papers*, No. 189, OECD Publishing, Paris.

Aspin, Christopher. 1964. "James Hargreaves and the Spinning Jenny." With Stanley D. Chapman. Helmshore Local History Society, Helmshore, Lancashire, U.K.

Brambrilla, Irene, Daniel Lederman, and Guido Porto. 2012. "Exports, Export Destinations, and Skills." *American Economic Review* 102 (7): 3406–38.

Brambrilla, Irene, and Darío Tortarolo. 2018. "Investment in ICT, Productivity, and Labor Demand: The Case of Argentina." Policy Research Working Paper 8325, World Bank, Washington, DC.

Chen, Maggie, and Min Xu. 2015. "Online International Trade in China." Background paper, World Bank, Washington, DC.

Dauth, Wolfgang, Sebastian Findeisen, Jens Südekum, and Nicole Wößner. 2017. "German Robots: The Impact of Industrial Robots on Workers." IAB Discussion Paper 30/2017, Institute for Employment Research, Nuremberg, Germany.

David, Benjamin. 2017. "Computer Technology and Probable Job Destructions in Japan: An Evaluation." *Journal of the Japanese and International Economies* 43 (March): 77–87.

de Soto, Hernando. 2002. *The Other Path: The Economic Answer to Terrorism.* New York: Basic Books.

eBay Inc. 2013. "Commerce 3.0 for Development: The Promise of the Global Empowerment Network." eBay Report Series, eBay Inc., Washington, DC.

Freund, Caroline L., Alen Mulabdic, and Michele Ruta. 2018. "Is 3D Printing a Threat to Global Trade?" Working paper, World Bank, Washington, DC.

Frey, Carl Benedikt, and Michael A. Osborne. 2017. "The Future of Employment: How Susceptible Are Jobs to Computerisation?" *Technological Forecasting and Social Change* 114(c): 254–80.

Glaeser, Edward L. 2018. "Framework for the Changing Nature of Work." Working paper, Harvard University, Cambridge, MA.

Gregory, Terry, Anna Salomons, and Ulrich Zierahn. 2016. "Racing with or against the Machine? Evidence from Europe." ZEW Discussion Paper 16–053, Center for European Economic Research, Mannheim, Germany.

Hallward-Driemeier, Mary, and Gaurav Nayyar. 2018. *Trouble in the Making? The Future of Manufacturing-Led Development.* Washington, DC: World Bank.

Hardy, Wojciech, Roma Keister, and Piotr Lewandowski. 2018. "Educational Upgrading, Structural Change, and the Task Composition of Jobs in Europe." *Economics of Transition* 26 (2): 201–31.

Keynes, John Maynard. [1930] 1963. "Economic Possibilities for Our Grandchildren." In *Essays in Persuasion*, 358–73. New York: W. W. Norton. http://www.econ.yale.edu/smith/econ116a/keynes1.pdf.

Mason, Andrew, Vera Kehayova, and Judy Yang. 2018. "Trade, Technology, Skills, and Jobs: Exploring the Road Ahead for Developing East Asia." Background paper, World Bank, Washington, DC.

McKinley, Richard Alexander, ed. 1958. *The City of Leicester.* Vol. 4 of *A History of the County of Leicester.* Victoria County History Series. Martlesham, Suffolk, U.K.: Boydell and Brewer.

TASS. 2017. "Sberbank Receives More than $2 Billion in Profits Each Year from the Introduction of Artificial Intelligence." September 25. http://special.tass.ru/ekonomika/4590924. In Russian.

Taylor, Kate. 2016. "Fast-Food CEO Says He's Investing in Machines because the Government Is Making It Difficult to Afford Employees." *Business Insider*, March 16.

Upwork. 2017. Freelancing in America 2017 (database). Mountain View, CA. https://www.upwork.com/i/freelancing-in-america/2017.

Verhoogen, Eric A. 2008. "Trade, Quality Upgrading, and Wage Inequality in the Mexican Manufacturing Sector." *Quarterly Journal of Economics* 123 (2): 489–530.

World Bank. 2014. "Vietnam Development Report 2014. Skilling Up Vietnam: Preparing the Workforce for a Modern Market Economy." World Bank, Washington, DC. http://documents.worldbank.org/curated/en/610301468176937722/pdf/829400AR0P13040Box0379879B00PUBLIC0.pdf.

———. 2015a. "Armenia: Skills toward Employment and Productivity (STEP), Survey Findings (Urban Areas)." World Bank, Washington, DC, January 31.

———. 2015b. "Georgia: Skills toward Employment and Productivity (STEP), Survey Findings (Urban Areas)." World Bank, Washington, DC, January 31.

———. 2016. *World Development Report 2016: Digital Dividends.* Washington, DC: World Bank.

Zeng Kunhua. 1973. *Zhongguo tie lu shi* [The history of Chinese railway development]. Vol. 1. Taipei, Taiwan, China: Wenhai Press.

CHAPTER 2

The changing nature of firms

Historically, firms have operated within boundaries. The British economist Ronald Coase explained this phenomenon in his 1937 article "The Nature of the Firm."[1] He observed that firms in Detroit grew only so long as it was cheaper for them to complete additional parts of the production process in-house rather than resort to the open market.

In 2018 firms operate within wider boundaries. Free trade agreements and improved infrastructure have reduced the cost of cross-border trade, allowing transactions to take place wherever costs are lower.[2] New technologies have lowered communication costs. As a result, firms are less vertically integrated—managers are outsourcing more tasks to the market. Some companies are even creating new markets. For example, JD.com, China's second-largest e-commerce company, has more than 170,000 online merchants on its platform, many in rural areas.

The wider boundaries of firms have evolved gradually. One only has to compare the Ford Motor Company of the 1930s with the Inter IKEA Group of 2018 to observe how firm boundaries have expanded over time. Henry Ford owned the sheep farms that supplied the wool for his company's automobile seat covers. He also owned the iron ore and coal freighters that fed the company's sprawling River Rouge manufacturing complex near Detroit. The company itself maintained most of the transactions related to car manufacturing in-house because the transaction costs were higher to find an outside supplier able to customize auto parts.

As for IKEA, vertical integration within Sweden gave ground to globalization in the 1980s and 1990s. The international expansion of IKEA, founded in Sweden in 1943, began with the establishment of small stores in Norway in 1963 and then in Denmark in 1969. The reduction in tariff and nontariff barriers allowed IKEA to set up global value chains. The advent of Internet technology transformed these chains into global networks: IKEA procures many of its products through online bidding. Firms from around the world become part of IKEA's network of suppliers.

Firms such as IKEA would have made Austrian economist Joseph Schumpeter proud. Capitalism requires "the perennial gale of creative destruction," wrote Schumpeter in 1942.[3] He did not worry about whether jobs might be lost in the process. Politicians do.

As the boundaries of firms have expanded, the corporate labor share has declined—in 75 percent of advanced countries and 59 percent of emerging economies between 1975 and 2012.[4] The World Bank, using total labor shares from Penn World Tables, including the self-employed and government sectors, has calculated a decline in two-thirds of 76 developing countries.

Governments are struggling to respond to this decline and often blame the rise of large firms to explain it. Politicians are trying to create jobs by financing programs for the development of small and medium enterprises. However, these programs are rarely cost-effective. They are based on the belief that small and medium enterprises create stable jobs, and yet the evidence shows that large firms account for the largest proportion of stable jobs in many economies.[5]

A better solution is to ease the barriers for start-ups to encourage competitive markets. Start-ups require a business-friendly environment that does not favor large private firms already in a market for some time (incumbents) and state-owned enterprises or firms run by government officials, their associates, or relatives. A small number of start-ups will become the next superstar firms.

Technological change favors the most productive firms in each industry, incentivizing the reallocation of resources toward them. Digital technologies allow for fast scaling. Jamalon, an online book retailer since 2010 in Amman, Jordan, has been able, with fewer than 100 staff, to establish partnerships with over 3,000 Arabic-language and 27,000 English-language publishers, delivering 10 million titles to the Middle East region. Platform-based businesses are on the rise across the globe, providing new opportunities to trade goods and services.

There is much to celebrate when it comes to the rise of large firms. The digital economy has enabled firms to grow faster than those 20 years ago. But there is also much to consider with caution. First, digital markets provide new opportunities for firms to stifle their competition. U.S. economist Sherwin Rosen, who introduced the concept of superstar firms in 1981, predicted that technology would allow firms to expand markets or crowd out the competition more easily. In many markets, this prediction has proven to be true. Technology has allowed some companies to rise quickly to the top, while preventing others from rising at all.

Second, globally integrated firms, which have "scale without mass,"[6] create new challenges for taxation. It is increasingly difficult to determine where value is created when businesses create value by combining networks of users, ideas, and production across borders. Taking advantage of international value creation, firms are in turn better able to divert profits to low-tax jurisdictions. Solutions require coordination at the global level. Meanwhile, countries could take unilateral steps by extending value added tax regimes or creating new taxes for the digital economy. But it is difficult to collect taxes on intangible assets such as user data. More traditional tax avoidance schemes, through transfer pricing, are also easier in the digital economy.

Superstar firms

Large firms dominate the global economy: 10 percent of the world's companies are estimated to generate 80 percent of all profits. Superstar firms shape a country's exports. One study of 32 developing countries found that, on average, the five largest exporters in a country account for a third of its exports, nearly half of export growth, and a third of growth due to export diversification.[7]

Growth is particularly strong in markets undergoing accelerated technological advances. Reduced trade barriers also encourage firms to grow by increasing access to new imported inputs.[8] The top 1 percent of exporters in rich countries account for a larger share of exports—on average, 55 percent—than those in poorer countries (figure 2.1).

FIGURE 2.1 **The top 1 percent of exporters account for a larger proportion of exports in rich countries than those in poorer countries**

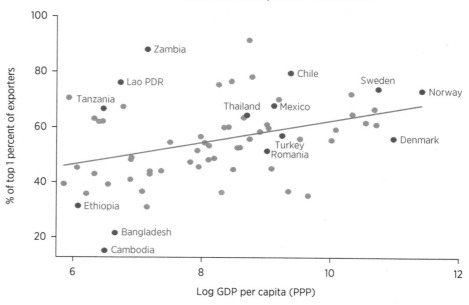

Source: WDR 2019 team, based on the Exporter Dynamics Database, version 2.0, described in Fernandes, Freund, and Pierola (2016).

Note: Oil exports (hydrocarbons such as oil, petroleum, natural gas, and coal) are excluded from the calculation. GDP = gross domestic product; PPP = purchasing power parity.

Large firms have a beneficial effect on economic growth. They accelerate growth in developing economies by pulling resources out of subsistence agriculture. Large firms are at the forefront of adopting new technologies. They increase aggregate productivity by upgrading their internal capabilities to become more efficient and drive out unproductive firms. And they achieve economies of scale that lower prices for consumers.

Large firms account for most formal jobs in an economy. In recent years, firms with more than 100 employees have accounted for 60 percent of the total employment share in Malaysia, Myanmar, and Vietnam. In Cambodia, they accounted for 70 percent. This situation plays out in other regions as well: large firms accounted for 53 percent of total employment in Argentina, 46 percent in Bolivia, 62 percent in the Dominican Republic, and 54 percent in Ecuador.[9] In Serbia, workers in the top 1 percent of manufacturing firms represented, on average, a quarter of total employment; the top 5 percent accounted for almost half of the total labor force.[10] The situation is similar in Romania. Superstar firms tend to employ the most workers because those firms generate the most output, even if they are less labor-intensive than the average firm.[11]

The importance of large firms in driving economic growth is not new. However, the advent of digital platforms has changed how this phenomenon unfolds. Digital platforms are replacing the brick-and-mortar malls, connecting shoppers with different brand stores, creating efficiencies for brands, and generating revenue for platform owners. Data gathered through platforms are also used to improve firm efficiency, at times in markets other than those in which the data were collected in the first place. JD Finance, a financial company that is part of the Chinese JD.com Group, incorporates transaction data that it gains through its marketplace into its loan assessment model.

Platform-based businesses are on the rise in every country. Consider VIPKID, a leading Chinese online education firm founded in 2013 that matches 500,000 students in China with 60,000 teachers in North America for virtual, one-on-one English classes. Jumia, established in 2012, is a Nigerian e-commerce company that has already made inroads into 23 African countries, bringing electronics, groceries, and fashion to customers. Flipkart in India facilitates sales of consumer electronics between suppliers and customers. It operates like a market, defying the boundaries of firms as originally described by Ronald Coase.

As noted, digital platforms allow quick scaling. Examples of billion-dollar start-ups built around platforms abound. In China, e-commerce giant JD.com started as a retail business in a tiny booth in the Zhongguancun Electronics Shopping Market in Beijing. In July 2018 the JD platform had 320 million active users. Ant Financial, part of the Alibaba Group, is the most valuable fintech firm in the world. The firm took off within just a few years because of advances in artificial intelligence. Ant uses big data to disburse loans in less than 1 second from the moment of application. Its "3-1-0" online lending model involves a 3-minute application process, a 1-second processing time, with 0 manual interventions. Since 2014, more than 4 million small Chinese businesses have received loans.

Digital platforms create instant business opportunities for entrepreneurs, thereby creating jobs. Since 2009, many clusters of rural micro e-tailers have opened shop on Taobao.com Marketplace, fostering "Taobao Villages" in China. Taobao Village merchants produce consumer goods, agricultural products, and handicraft works based on their niche competencies. Taobao Villages have created more than 1.3 million jobs, drawing young people who migrated to cities back to their hometowns to start their own enterprises. Reliable Internet connectivity and high smartphone penetration must be in place for this kind of e-commerce to grow.

Platforms expand job opportunities. In 2018 the service sector accounted for most jobs in several countries: its share of total employment was more than 70 percent in Argentina, Saudi Arabia, and Uruguay, and above 80 percent in Hong Kong SAR, China; Israel; and Jordan. The proliferation of platforms is allowing freelancers to have simultaneous access to multiple platforms at low entry costs. Consumers are also more willing to use online services because they trust them, relying on brand certification, digitalized social capital, and third-party validations. Consumer trust enables platforms to expand quickly into other business lines. Grab, a Singapore-based ride-hailing platform, eventually captured 95 percent of the Southeast Asian market before expanding to offer additional services ranging from food ordering to payment systems. GrabPay extends e-payment opportunities to an estimated two-thirds of people in the region who do not have bank accounts.

Some platforms expand the supply of labor by increasing opportunities for new, flexible types of work that complement traditional forms of employment in the gig economy. Workers set their own hours for most platforms. Meanwhile, the additional income may reduce income fluctuations for secondary earners. The flexibility inherent in platform work also enables more women to participate in the labor force. But these features blur the line between formal and casual employment. Although flexibility is a benefit in some cases, it also raises concerns around income instability and protections connected with standard employer-employee relationships, including pension plans, health insurance, and paid leave.

Finally, digital platforms enable firms to exploit underused physical and human capacity, transforming dead capital into active capital. For example, ride-hailing platforms provide a way for individuals to advertise their free time and spare vehicle capacity—be it a luxury vehicle, a moped, or a tuk-tuk—to generate income. Freelancing websites enable unemployed computer programmers located in remote parts of the world to document their expertise so they can find work with companies abroad.

The rise of the digital platform firm—operating globally, existing principally in the cloud, and often generating income from the capital of others—marks a shift in the potential nature of firms more generally. Most regulations are not yet adapted to these changes. Platform firms often operate in regulatory gray areas, but minimum standards of quality, prudence, and safety, among other policy goals, must still be upheld by digital businesses. Data privacy and protection are at the center of the regulatory discussion because of the large amount of data accumulated, employed, and monetized by platform businesses. Zoning or other laws affecting business activity may also be implicated. For example, although Airbnb frequently shifts tourism away from urban centers and has a positive impact on local businesses, Airbnb locations are often not subject to the same zoning or licensing requirements as other commercial accommodations. Nevertheless, Airbnb might affect neighbors who do not share the benefit of rental income.

Regulation becomes important if platforms provoke a race to the bottom in working conditions. In Indonesia, drivers with Go-Jek and Grab held large demonstrations in early 2018 demanding an increase in their tariffs. In response, the Indonesian government is amending its laws to require such firms to register as transport companies, comply with safety requirements, and impose a minimum floor price. In early 2018, Egyptian courts suspended the licenses of the ride-hailing companies Uber and Careem in response to a challenge by taxi drivers. Shortly thereafter, in May 2018, the Egyptian government passed a law to regulate ride-hailing companies, allowing Uber and Careem to get back on the road and compete alongside traditional taxis.

Competitive markets

Physical presence is no longer a prerequisite to doing business in a given market, particularly in the digital economy where intangible products are replicable at little or no cost. Expanded boundaries create opportunities for firms to grow, but frequently the risk of market concentration increases. Anticompetitive behavior is harder to identify in the digital economy. Network effects often benefit early adopters of technology, facilitating the emergence of monopolies.

More start-ups mean more competition. If the business conditions are right, it is more likely that some start-ups will grow strongly, creating jobs. Faced with new competition, less productive firms—so long as they are not state-owned or politically connected—exit the market.[12]

Creating a better business environment allows more successful firms to rise naturally. The World Bank's Doing Business project lays out the basic regulatory requirements for private initiatives to grow. These data have been used by researchers to study the deleterious effects of burdensome regulation. Poverty rates are lower in countries that have business-friendly regulations.[13]

A country with a business-friendly environment also has more start-up activity and job creation. When Mexico simplified business registration, the number of new businesses increased by 5 percent, and wage employment went up by 2.2 percent.[14] Higher start-up costs may also lead to lower overall productivity: in the absence of competition, firms that are already in a market will continue to operate regardless of productivity levels.

Competitive markets and efficient trade require basic infrastructure—roads, bridges, ports, and airports. Lower transport costs, as well as streamlined, cheaper border compliance processes, increase exports. Logistics infrastructure facilitates the online trading of nondigital products.

Broadband access is a prerequisite for business in the digital era—after all, many firms depend in part or even exclusively on the Internet. Mobile phone access alone is not enough; broadband technologies push down transaction costs even further in remote markets that lack transport infrastructure. Those living in the Middle East and North Africa region are some of the most underserved. Although that region can boast over 120 mobile phone subscriptions for every 100 inhabitants (one of the highest levels in the world), it has fewer than 10 broadband subscriptions per 100 inhabitants, and bandwidth per subscriber is limited. In the end, this means that although the citizens of these countries are active on social media, digital finance has barely any presence.

Technology allows firms to sharpen their competitive edge by making their operations more efficient and by enabling them to create new ways of doing business. Teleroute, a Belgian platform, uses an algorithm to match freight forwarders and carriers in Europe, reducing empty runs by up to 25 percent. Improved connectivity also enables start-ups to source essential technical expertise through online freelancing platforms. Upwork, a U.S.

platform, has since 2015 connected 5 million businesses with more than 12 million freelancers. Its fourth-largest community of task providers is in Ukraine. At one time, start-ups required data centers, information technology systems, custom software, and a user support infrastructure to take on large conglomerates. In the digital age, entrepreneurs worldwide have access to these services via the Internet.

One aspect of the digital economy is that it poses new challenges for competition law, mergers and acquisitions, and consumer welfare. The ascent of platform firms raises issues related to market power (figure 2.2). The network effects associated with some online products often lead to significant benefits for early adopters, resulting in market concentration and facilitating the emergence of monopolies. In 2017 Safaricom, Kenya's largest mobile phone operator, with 80 percent of the market, launched M-Pesa, the country's first mobile money system. A year later, M-Pesa commanded the same market share in mobile money.

Platform firms at times exclude competitors by charging higher fees for other networks to interconnect. When Zimbabwe mandated interoperability and infrastructure sharing among e-money operators, the number of subscribers rose by 15 percent. In Peru, the telecom regulator forced the largest communication networks to offer messaging services to banks that were expanding into e-money.

Overall, the digital economy poses challenges for policy makers. Many digital platform companies operate in adjacent, multisided markets, bundling or at least connecting different types of services. New types of market power emerge when firms provide services free of charge on one side of the market in exchange for user data and then they monetize that data on another side of the market. Antitrust investigations are having to adjust to these new situations and use new rules of analysis.

Tax avoidance

With the boundaries of firms transcending borders and physical assets, it has become easier to shift profits to low-tax jurisdictions (tax planning and tax avoidance). As a result, billions of dollars of corporate profits go untaxed every year. The international tax system would benefit from an update; how to tax

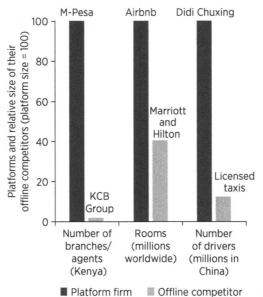

FIGURE 2.2 Platform firms dominate their offline competitors in 2018

Sources: WDR 2019 team based on data from Safaricom; KCB Bank Group; Airbnb; Marriott International Inc.; *Financial Times.*

businesses in the globalized digital economy and how to distribute value are questions under debate.

The Organisation for Economic Co-operation and Development (OECD) estimates that US$100–$240 billion is lost in revenue each year from base erosion and profit shifting by multinational companies.[15] This amount is equivalent to 4–10 percent of global corporate income tax revenue. Another estimate suggests that multinationals shift close to 40 percent of their profits to tax havens, resulting in a loss of 12 percent of global corporate tax revenues.[16] These losses are not equal across countries, however. Australia, Brazil, France, India, Japan, Mexico, and the United States, as well as much of Africa, are among the countries estimated to be most hurt by profit shifting.[17]

A multitude of loopholes can be found in tax laws, many created through corporate lobbying, that enable companies to reduce their tax burden. These loopholes can be used by corporations to increase their tax deductions and move profits elsewhere, at times to jurisdictions with low or zero corporate income taxes—so-called tax havens or investment hubs. This phenomenon is neither new nor illegal, but it is easier in the digital economy. In 2016 almost 60 percent of Fortune 500 companies had at least one affiliate established in Bermuda or the Cayman Islands, both of which have a 0 percent corporate income tax rate. The Paradise Papers, leaked in late 2017, are exposing many examples of tax avoidance and tax evasion worldwide.[18]

The problem is that the current rules are based on both source and residence. Source relates to a justification based on the geographic location of the income-generating activities (the idea of "where value is created" linked to the physical presence of labor or capital). Residence refers to where the company receiving the income is considered to have its primary location, usually based on where the company is incorporated or effectively managed as per the owner's linkage to the state (residence, domicile, or citizenship—physical presence). Source countries have primary taxing rights over the income from sales. Residence countries tax multinationals' income from cash investments.

In practice, under the prevailing rules multinational enterprises pay taxes in the countries in which they locate their affiliates and activities. Firms organize their own internal cross-border production structures between affiliates, declaring different profits for different affiliates, in some cases seemingly irrespective of the direct value generation by each affiliate. It is often difficult to identify when these structures are legitimate and when they are established principally to avoid paying taxes in higher-tax jurisdictions.

Because they have many opportunities to avoid paying taxes, it is not surprising that firms do so. In fact, profits have become more sensitive to international tax differentials over time, which means firms are getting better at avoiding taxes. Some firms engage in transfer mispricing, charging lower prices for exports sold from high-tax countries to low-tax countries, or higher prices for inputs coming from low-tax countries. The strategic location of intellectual property ownership, international debt shifting through intracompany loans, treaty shopping, and tax deferrals are other mechanisms used to avoid taxes. Effective corporate taxation rates have a decisive impact on where affiliates locate. A 1 percentage point larger tax rate differential between two jurisdictions reduces the reported pretax profits of an affiliate by 1 percent.[19] Tax treaties are estimated to have reduced tax revenues in Africa by about 8.5 percent among those countries that signed a treaty with an investment hub.[20]

The digital economy poses new challenges. The virtual nature of digital businesses makes it even easier to locate activities in low-tax jurisdictions. The provision of goods and services from abroad without a physical presence in countries where consumers are located escapes the traditional corporate tax. Digital firms are able to generate profit out of intangible assets such as (foreign) user data. As a result, identifying where value is created is difficult.

In 2016 OECD released a template for collecting the value added tax from foreign suppliers of digital goods and services. Since then, more than 50 countries have adopted the recommended guidelines for imposing the tax on the direct consumer sales of services and intangibles by foreign suppliers. More recently, in early 2018, the OECD Task Force on the Digital Economy (comprising over 110 countries and jurisdictions) released an interim report on tax challenges arising from digitization and has committed to delivering a long-term, consensus-based solution by 2020.

The European Union has levied the value added tax on nonresident suppliers of telecommunications, broadcasting, and electronic services, regardless of scale, since January 2015. Nonresident businesses are required to charge the customer the tax at the rate applying in the customer's country, removing the competitive advantage held by digital companies located in countries with low value added tax rates. This new value added tax has raised more than €3 billion for the European Union. Australia adopted a similar approach in July 2017. Singapore announced in its February 2018 budget that a goods and services tax will be imposed on imported services, including digital services such as music and movie streaming, starting in January 2020. Other advanced economies with indirect taxes on the digital economy are Japan, the Republic of Korea, New Zealand, Norway, the Russian Federation, and South Africa.

Less has been done in emerging economies, where the additional tax revenues are needed the most. In 2017 Serbia and Taiwan, China, adopted models extending their value added tax regimes to cover digital suppliers. In 2018 Argentina and Turkey did the same. China, Malaysia, and Thailand are among the countries reviewing their tax laws to extend collection to digital profits.

As an alternative, a government could introduce a new freestanding tax on foreign suppliers of digital services. Such a tax would do a better job of directly targeting foreign suppliers rather than domestic consumers. It would avoid conflict with existing double taxation agreements by being separate from the mainstream income tax system. Arguably, a freestanding tax would level the playing field between domestic and foreign suppliers of digital services. As for the value added tax, collecting this kind of tax is enhanced through a registry of nonresident suppliers of digital services.

In 2016 the government of India introduced a 6 percent equalization levy on online advertising revenue paid by Indian companies to nonresident e-commerce companies. In March 2018 the European Commission proposed a tax on the gross revenues from digital activities in which users have a major role in value creation. The tax would apply to revenues from selling online advertising space, intermediary activities that allow users to interact and sell goods and services, and the sale of data. The Commission has estimated that a 3 percent tax could raise €5 billion a year. European Union members have yet to reach a consensus on whether to adopt the tax.

Along with the adoption of new measures to tax digital business, the international community has been taking steps to address base erosion and profit shifting, as well as other tax avoidance or evasion schemes, by both digital and traditional business. The Global Forum on Transparency and Exchange of Information for Tax Purposes brings together almost 150 jurisdictions to implement internationally agreed standards on transparency and the exchange of information for tax purposes. In addition, the Base Erosion and Profit Shifting initiative launched by OECD and G-20 countries in 2013 brings together more than 115 countries to reduce tax avoidance. The group negotiated a comprehensive package of measures to reduce profit shifting, placing more emphasis on the source principle. Emphasizing the source principle better aligns the location of taxable profits with the location of value creation and improves the information available to tax authorities.

In addition to these initiatives, other steps to strengthen tax administration include widening the definition of a "permanent establishment" to recognize that companies may conduct considerable business in a country without having much of a physical presence; strengthening transfer pricing and anti–tax avoidance rules and audit capacity by emerging economies; adopting some aspects of formulary apportionment; and relying on targeted anti–tax avoidance measures such as stronger controlled foreign corporation rules.

Governments are taking some steps unilaterally. New anti-diversion rules entered into force in the United Kingdom in 2015, allowing firms to pay taxes up-front. The rules are designed to incentivize greater compliance with the mainstream corporate income tax regime. Australia adopted similar rules in 2016. Meanwhile, these countries already have a transfer pricing capacity and a sophisticated suite of anti–tax avoidance measures. The context for emerging economies is rather different, however. Their capacity to address transfer pricing risks is low, so anti–tax avoidance legislation may not be effective. By integrating a more mechanical and targeted anti-diversion rule into their corporate tax systems, these countries could better tackle tax avoidance. Certain minimum criteria as set out in the law could trigger the application of such a rule. Some countries also impose a minimum tax based on turnover.

Growing public discontent with tax avoidance practices by multinational firms has revived discussions around more significant overhauls of the international tax system. Global formulary apportionment is one option that has garnered much interest among some policy makers, although the challenges involved in reaching global agreement and implementation make it unlikely to be adopted. Such a system would divide the tax base between jurisdictions according to the location in which source activities take place.

Governments would have to agree on a formula to allocate profits, usually based on tangible assets such as volume of sales to third parties, assets, payroll, or a head count of staff in each jurisdiction. This system is used domestically in countries such as Canada and Switzerland to apportion income among provinces and cantons. Although global formulary apportionment would remove existing incentives for profit shifting to low-tax jurisdictions, it would create incentives for other methods of profit shifting.

Another option is a destination-based cash flow tax (or border-adjusted tax), which is similar to formulary apportionment based exclusively on volume of sales, but the tax base is not consolidated. Instead, governments tax net income from sales in the purchaser's place of residence.

The international community continues to take steps to mitigate weaknesses in global corporate tax. Important progress is being made in adjusting tax systems to the new nature of firms, particularly platform companies, but much more work remains to be done.

Notes

1. Coase (1937).
2. Djankov, Freund, and Pham (2010).
3. Schumpeter ([1942] 2003, 84).
4. Karabarbounis and Neiman (2013).
5. Freund (2016).
6. Brynjolfsson et al. (2008).
7. Freund and Pierola (2015).
8. Goldberg et al. (2010).
9. All figures are based on 2014–17 data from the World Bank's Enterprise Surveys.
10. Figures are based on the Exporter Dynamics Database, version 2.0, with additional data updates.
11. Freund (2016).
12. Rijkers, Freund, and Nucifora (2017).
13. Djankov, Georgieva, and Ramalho (2018).
14. Bruhn (2011).
15. OECD (2017).
16. Tørsløv, Wier, and Zucman (2018).
17. Beer, de Mooij, and Liu (2018).
18. The Paradise Papers refers to a set of 13.4 million confidential electronic documents related to offshore investments that were leaked to the German media in late 2017. See the website of the International Consortium of Investigative Journalists for a database of the documents (http://www.icij.org /investigations/paradise-papers).
19. Beer, de Mooij, and Liu (2018).
20. Beer and Loeprick (2018).

References

Beer, Sebastian, Ruud de Mooij, and Li Liu. 2018. "International Corporate Tax Avoidance: A Review of the Channels, Magnitudes, and Blind Spots." IMF Working Paper WP/18/168, International Monetary Fund, Washington, DC, July 23.

Beer, Sebastian, and Jan Loeprick. 2018. "The Costs and Benefits of Tax Treaties with Investment Hubs: Findings from Sub-Saharan Africa." Working paper, International Monetary Fund, Washington, DC.

Bruhn, Miriam. 2011. "License to Sell: The Effect of Business Registration Reform on Entrepreneurial Activity in Mexico." *Review of Economics and Statistics* 93 (1): 382–86.

Brynjolfsson, Erik, Andrew McAfee, Michael Sorell, and Feng Zhu. 2008. "Scale without Mass: Business Process Replication and Industry Dynamics." Harvard Business School Technology and Operations Management Unit Research Paper No. 07-016, Cambridge, MA.

Coase, Ronald. 1937. "The Nature of the Firm." *Economica* 4 (16): 386–405.

Djankov, Simeon, Caroline L. Freund, and Cong S. Pham. 2010. "Trading on Time." *Review of Economics and Statistics* 92 (1): 166–73.

Djankov, Simeon, Dorina Georgieva, and Rita Ramalho. 2018. "Business Regulations and Poverty." *Economics Letters* 165 (April): 82–87.

Fernandes, Ana M., Caroline L. Freund, and Martha Denisse Pierola. 2016. "Exporter Behavior, Country Size and Stage of Development: Evidence from the Exporter Dynamics Database." *Journal of Development Economics* 119(C): 121–37.

Freund, Caroline L. 2016. *Rich People, Poor Countries: The Rise of Emerging-Market Tycoons and Their Mega Firms.* Assisted by Sarah Oliver. Washington, DC: Peterson Institute for International Economics.

Freund, Caroline L., and Martha Denisse Pierola. 2015. "Export Superstars." *Review of Economics and Statistics* 97 (5): 1023–32.

Goldberg, Pinelopi Koujianou, Amit Kumar Khandelwal, Nina Pavcnik, and Petia Topalova. 2010. "Imported Intermediate Inputs and Domestic Product Growth: Evidence from India." *Quarterly Journal of Economics* 125 (4): 1727–67.

Karabarbounis, Loukas, and Brent Neiman. 2013. "The Global Decline of the Labor Share." *Quarterly Journal of Economics* 129 (1): 61–103.

OECD (Organisation for Economic Co-operation and Development). 2017. "Background Brief: Inclusive Framework on BEPS." Paris, January. http://www.oecd.org/tax/beps/background-brief-inclusive-framework-for-beps-implementation.pdf.

Rijkers, Bob, Caroline L. Freund, and Antonio Nucifora. 2017. "All in the Family: State Capture in Tunisia." *Journal of Development Economics* 124 (January): 41–59.

Schumpeter, Joseph Alois. [1942] 2003. *Capitalism, Socialism, and Democracy.* London: Routledge.

Tørsløv, Thomas, Ludvig Wier, and Gabriel Zucman. 2018. "The Missing Profits of Nations." Working paper, University of California, Berkeley. http://gabriel-zucman.eu/missingprofits/.

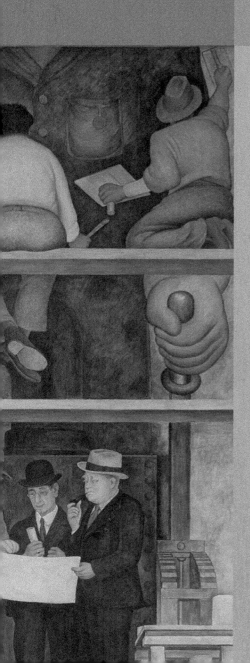

CHAPTER 3

Building human capital

The world is healthier and more educated than ever. In 1980 only 5 in 10 primary school-age children in low-income countries were enrolled in school. By 2015 this number had increased to 8 in 10. In 1980 only 84 of 100 children reached their fifth birthday, compared with 94 of 100 in 2018. A child born in the developing world in 1980 could expect to live for 52 years. In 2018 this number was 65 years.

But a large and unfinished agenda remains. Life expectancy in the developing world still lags far behind that of rich countries such as the Republic of Korea, where a girl born in 2018 can expect to live more than 85 years. Nearly a quarter of children under age 5 are malnourished. In many places, the working memory and executive functions (such as sustained attention) of poor children begin to lag as early as at 6 months of age.[1] Worldwide, more than 260 million children and youth are not in school. Meanwhile, nearly 60 percent of primary school children in developing countries fail to achieve minimum proficiency in learning.

Human capital consists of the knowledge, skills, and health that people accumulate over their lives, enabling them to realize their potential as productive members of society. It has large payoffs for individuals, societies, and countries. This was true in the 1700s when the Scottish economist Adam Smith wrote, "The acquisition of . . . talents during . . . education, study or apprenticeship, costs a real expense, which [is] capital in [a] person. Those talents [are] part of his fortune [and] likewise that of society."[2] This is still true in 2018.

For individuals, an additional year of school generates higher earnings on average. These returns are large in low- and middle-income countries, especially for women. However, what children learn matters more than how long they stay in school. In the United States, replacing a low-quality teacher in an elementary school classroom with an average-quality teacher raises the combined lifetime income of that classroom's students by US$250,000.[3]

Despite the larger supply of educated workers, returns to investments in education have increased since 2000.[4] Returns to education are especially high when technology is changing—people with higher human capital adapt faster to technological change. Indeed, a worker's future success depends on working with machines, not fearing them. In Mexico, the benefits of increased labor productivity resulting from the 1994 North American Free Trade Agreement (NAFTA) have been concentrated among more skilled workers.

Developing sociobehavioral skills such as an aptitude for teamwork, empathy, conflict resolution, and relationship management enlarges a person's human capital. Globalized and automated economies put a higher premium on human capabilities that cannot be fully mimicked by machines. Abilities such as grit have economic returns that are often as large as those associated with cognitive skills.

Health is an important component of human capital. People are more productive when they are healthier. In Nigeria, a program providing malaria testing and treatment increased workers' earnings by 10 percent in just a few weeks.[5] A study in Kenya showed that deworming in childhood reduced school absences while raising wages in adulthood by as much as 20 percent, all thanks to a pill that costs 25 cents to produce and deliver.[6]

From an early age, the dimensions of human capital complement each other. Proper nutrition in utero and in early childhood improves children's physical and mental well-being. Evidence from the United Kingdom revealed that schoolchildren who had healthier diets significantly increased their achievements in English and science.[7] Meanwhile, a multicountry study in Southeast Asia found that both underweight and obese children had lower IQ scores than healthy-weight children.[8] In India, giving preschoolers mathematics-based games generated enduring improvements in their intuitive abilities.[9]

The benefits of human capital transcend private returns, extending to others and across generations.[10] Deworming one child decreases the chances of other children becoming infected with worms, which in turn sets those children up for better learning and higher wages.[11] Maternal education, through better prenatal care, improves infant health. In Pakistan, children whose mothers have even a single year of education spend an extra hour a day studying at home.[12]

These individual returns to human capital add up to large benefits for economies—countries become richer as more human capital accumulates. Human capital complements physical capital in the production process and is an important input to technological innovation and long-run growth. As a result, between 10 and 30 percent of per capita gross domestic product (GDP) differences is attributable to cross-country differences in human capital.[13] This percentage could be even higher when considering the quality of education or the interactions between workers with different skills. And not to be overlooked, by generating higher incomes, human capital accelerates the demographic transition and reduces poverty.

Over the longer term, human capital matters for societies. In the mid-1970s, Nigeria introduced universal primary education, sending a large cohort of children through primary school who otherwise would not have gone. Years later, the members of that cohort were found to be more engaged in political life. They paid closer attention to the news, spoke to their peers about politics, attended community meetings, and voted more often than those who did not go to primary school. Young participants in the National Volunteer Service Program in Lebanon, an intercommunity soft skills training program, display higher levels of overall tolerance. As the scientist Marie Curie once said, "You cannot hope to build a better world without improving the individuals."

Human capital also fosters social capital. Surveys typically find that more educated people are more trusting of others. Research suggests that the large wave of compulsory school reforms that took place across Europe in the mid-20th century made people more tolerant of immigrants than they were before.[14] Social capital in turn is associated with higher economic growth.[15] Conversely, failing to protect human capital undermines social cohesion.

Human capital is one of the first things to suffer when things fall apart. Wars often prevent whole generations from realizing their potential. For example, between 2011 and 2017 almost 4 million Syrian children left school because of the civil war. Many of them are likely to never make up for these lost years of school (figure 3.1).

FIGURE 3.1 In the Syrian Arab Republic, the number of children out of school because of war rose between 2011 and 2017

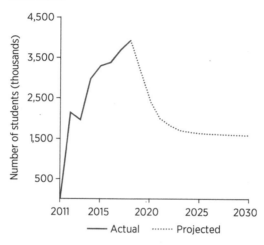

—— Actual ········ Projected

Source: WDR 2019 team.

Note: The number of children out of school between 2011 and 2017 is based on estimates of actual declines in school enrollment relative to prewar trends and on the assumed impact war has had on student enrollment. The scenario from 2018 onward explores the long-term consequences of these trends by assuming that school enrollments gradually return to prewar trends and corrects for the population dynamics of refugee in-and-out flow rates (if they are similar to those after past international conflicts). Similar assumptions are also made for internally displaced persons but with higher return rates during the first few years after the end of the war.

Why governments should get involved

Individuals and families often cannot afford the costs of acquiring human capital. Even when human capital investments are affordable, individual decisions may be shaped by lack of information, or restricted because of the prevalent social norms. Individuals also do not necessarily consider the wider social benefits for others. For these reasons, governments have an important role to play in fostering human capital acquisition.

Many disadvantaged families want to invest in better health and education for their children but cannot afford to do so. The proof is how families spend their money once budget constraints are even slightly relaxed. In Sierra Leone, only three to four months after the introduction of a public works program that increased income, participating families significantly increased their spending on health services, especially for children.[16]

Even when education is free, the cost of transportation and school supplies, together with the earnings lost while a child is in school instead of working, make education prohibitively expensive. Many poor rural families cannot afford the time it takes to travel to the nearest school or medical facility. In Niger, only 24 percent of the population lives within a one-hour walk of the nearest medical facility during the wet season.[17]

In cases such as these, government interventions make a big difference. Cash transfer programs have improved the health and education of millions of children in low- and middle-income countries, even when they have provided only partial subsidies for the cost of school. Shombhob, a conditional cash transfer program piloted in Bangladesh, reduced wasting among children ages 10–22 months and taught mothers about the benefits of breastfeeding.[18] And the effects of these programs are felt over time. A two-year conditional cash transfer program in Malawi targeting adolescent girls and young women produced a large increase in educational attainment and a sustained reduction in the total number of births in girls who were out of school at the start of the program. These benefits persisted after the program ended.[19]

Programs can improve people's incentives to invest in human capital when they make its long-term benefits salient or provide mechanisms to make good choices binding. Young people may not want to stay in school or take care of their health because they lack self-control or do not fully appreciate the benefits of education and good health.[20] However, when they receive information about human capital it has big effects on their behavior. In the Philippines, young people were offered a voluntary commitment program in which money they had placed in a savings account was returned to them only if they passed a smoking cessation test. The program saw a significant reduction in smoking.[21]

Human capital investment generates significant social returns as well, but these are often hard for parents to quantify, let alone factor into their decisions. When deciding to deworm their children, parents may not consider the fact that other children are also less likely to be infected. Parents deciding to send their children to preschool may not consider the wider future societal benefits such as lower crime and incarceration rates that have been associated with early childhood development programs. A 2010 study of Perry Preschool, a high-quality program for 3- to 5-year-olds developed in the 1960s in Michigan in the United States, estimated a return to society over and above the private return of about US$7–$12 for each dollar invested.[22] Without government interventions, families might not choose to invest enough in these types of programs.

Ensuring access to a quality education closes early gaps in cognitive and sociobehavioral skills. By the age of 3, children from low-income families have heard 30 million fewer words than their more affluent peers. As children turn into teenagers, interventions to close these gaps become more expensive. Evidence suggests that, for governments seeking to invest wisely in human capital, there is no better possibility than investing in the first thousand days of a child's life. Without such interventions early in life, it is more likely that a spiral of increasing inequality will ensue: subsequent public investments in education and health are more likely to benefit people who start out better off.

Government actions to support investment in human capital go well beyond spending on health, education, and social protection programs. In Nepal, investments in sanitation are contributing significantly to preventing anemia.[23] Housing programs improve the education and labor market outcomes of the most disadvantaged by changing the quality of the peers with whom they interact. The earlier children are exposed to better-off neighbors, the stronger are the effects.

Why measurement helps

Governments have a vital role to play in building human capital: as providers of health, education, and financing to ensure equitable access to opportunities and as regulators of accreditation and quality control of private providers. And yet they often fail to deliver. Most governments commit a significant share of their budgets to education and health, but public services are often too low quality to generate human capital. Sometimes, those services fail only the poor. Sometimes, they fail everyone—and the rich simply opt out of the public system.

Shortfalls in quality persist for two reasons. First, pursuing good policies does not always pay off politically. Second, bureaucracies may lack the capacity or incentives to convert good policies into effective programs. If public health is not politically relevant until there is a health crisis, politicians have little reason to prepare for future pandemics. Even when politicians and voters agree on the importance of an issue, they may disagree about the solution. Rarely is it popular to fund public health programs by raising taxes or by diverting money from more visible expenditures, such as on infrastructure or public subsidies.

The government of Nigeria encountered resistance in 2012 when it tried to repeal fuel subsidies to spend more on maternal and child health services. The media focused on the unpopular subsidy repeal and paid scant attention to the much-needed expansion of primary health care. The subsidy was thus reinstated because of public protests. Such a response to proposed changes occurs in some countries because the organized interests that stand to lose from reforms are powerful. In others, it happens because of a weak social contract: citizens do not trust their government, and so they are hesitant to pay taxes that they worry will be misspent. The consequence is that governments favor spending more on the politically visible aspects of human capital such as constructing schools and hospitals but much less on intangible aspects—such as the quality and competence of teachers and health workers. Campaigning politicians often promise new schools or hospitals, but rarely do they discuss actual learning levels or stunting rates.

Because investments in human capital may not produce economic returns for years, politicians tend to think of shorter-term ways to burnish their reputations. Although people with a basic education earn more than people with no education, labor market returns for a basic education are not realized until 10–15 years after these investments are made. This is even truer of investments in early childhood education. In Jamaica, providing toddlers with psychosocial stimulation increased earnings by 25 percent, but these returns only materialized 20 years later.[24]

One illustration of how technical and political complexities get in the way of delivering human capital interventions is in the area of early childhood development. Scholars generally agree that investments in children have high rates of return. However, challenges make the large-scale implementation of such investments difficult. First, as just noted, it takes a long time for society to benefit from these investments. Second, services have to be delivered in a synergic way over a short period within a person's life cycle. Third, multiple government departments are involved in the delivery of early childhood investments. Still, the experiences of countries such as Brazil, Chile, and Colombia reveal that large-scale early childhood development policies are feasible. One program, Chile Crece Contigo (Chile Grows with You), launched in 2006, serves as a reference point for middle-income countries willing to invest in children on a large scale. The Chilean early childhood development program integrates health, education, and social protection services for young children, combining universal and targeted programs. Rigorous evaluations boost the demand for political commitment.

Bureaucracies charged with implementing policies to build human capital often lack the capacity or the incentives to do so effectively. The World Bank's Service Delivery Indicators surveys conducted in seven countries in Sub-Saharan Africa (together representing close to 40 percent of the continent's population) found that, on average, 3 in 10 fourth-grade teachers had not mastered the language curriculum they were teaching. On a positive note, 94 percent of Kenyan teachers had done so. The surveys paint an equally mixed picture for health care facilities: about 80 percent of Kenyan doctors could correctly diagnose a basic condition such as neonatal asphyxia, whereas less than 50 percent of Nigerian doctors were able to do so.

Better measurement of outcomes sheds light on the political and bureaucratic failures that lead to the poor-quality delivery of social services. Information is an essential first step toward encouraging citizens to demand more from their leaders and service providers. In Uganda, releasing report cards on the performance of local health facilities galvanized communities to press for service delivery reforms. This pressure led in turn to sustained improvements in health outcomes, including a reduction in mortality for children under 5.

Better measurement also increases policy makers' awareness of the importance of investing in human capital, thereby creating momentum for action. Twaweza, a Tanzanian organization, launched a survey to assess children's basic literacy and numeracy. The dismal results—released in 2011—showed that only 3 in 10 third-grade students had mastered second-grade numeracy, and even fewer could read a second-grade story. The World Bank's own Service Delivery Indicators, released around the same time, shone a spotlight on the low levels of teacher competence and high levels of absenteeism in Tanzania. Together, these results led to a loud public outcry and the introduction of Tanzania's Big Results Now initiative, a government effort to track and address low levels of learning. It is already leading to tangible results.

More information is also needed to design and deliver cost-effective policies, even when a government is fully willing to invest in human capital. Both Peru and Vietnam have implemented ambitious policies to improve human capital. But only a comprehensive measurement of the factors that contribute to individual learning will shed light on the reasons behind differentials between these two countries. Once the gaps have been identified, cost-effective policies have to be designed and brought to scale.

The human capital project

Credible measurement of education and health outcomes raises the importance of human capital locally, nationally, and globally. Measurement spurs the demand for policy interventions to build human capital in countries where governments are not doing enough. Good measurement is essential to developing research and analysis to inform the design of policies that improve human capital.

With this goal in mind, the World Bank has launched the human capital project—a program of advocacy, measurement, and analytical work to raise awareness and increase demand for interventions to build human capital. The project has three components: (1) a cross-country metric—the human capital index, (2) a program of measurement and research to inform policy action, and (3) a program of support for country strategies to accelerate investment in human capital.

The first step in the project is an international metric to benchmark certain components of human capital across countries.[25] The new index measures the amount of human capital that a child born in 2018 can expect to attain by age 18 in view of the risks of poor education and poor health that prevail in the country in which she was born. The index is designed to highlight how improvements in the current education and health outcomes shape the productivity of the next generation of workers: it assumes that children born in a given year experience current educational opportunities and health risks over the next 18 years. A focus on outcomes—and not inputs such as spending or regulation—directs attention to results, which are what really matter. It also makes the human capital index relevant to the policy makers who design and implement interventions to improve these outcomes in the medium term.

The index follows the trajectory from birth to adulthood of a child born in a given year. In the poorest countries, there is a significant risk that the child does not even survive to see her fifth birthday. Even if she does reach school age, there is a further risk that she does not start school, let alone complete the full cycle of education through grade 12 that is the norm in rich countries. The time she does spend in school may translate unevenly into learning, depending on the quality of her teachers and schools and the support she receives from her family. After she reaches her 18th year, she carries with her the lasting childhood effects of poor health and nutrition that limit her physical and cognitive abilities as an adult.

The human capital index quantifies the milestones in this trajectory in terms of their consequences for the productivity of the next generation of workers. It has three components: (1) a measure of whether children survive from birth to school age (age 5); (2) a measure of expected years of quality-adjusted school, which combines information on the quantity and quality of education (figure 3.2, panel a); and (3) two broad measures of health—stunting rates (figure 3.2, panel b) and adult survival rates.

Survival to age 5 is measured using under-5 mortality rates compiled by the United Nations Inter-agency Group for Child Mortality Estimation. Nearly all children survive from birth to school age in rich countries. But in the poorest countries, as many as 1 in 10 children do not see their fifth birthday. The deaths of these children are not just a tragedy, but also a loss of their human capital, which never is realized.

FIGURE 3.2 Learning and stunting are two components of the human capital index

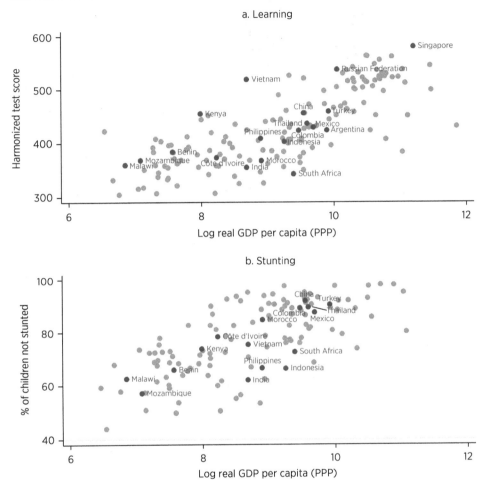

a. Learning

b. Stunting

Sources: Harmonized test scores are drawn from Patrinos and Angrist (2018); stunting data are from the UNICEF–WHO–World Bank Joint Child Malnutrition Estimates database, supplemented with data provided by World Bank country teams.

Note: GDP = gross domestic product; PPP = purchasing power parity.

The quantity of education is measured as the number of years of school a child can expect to obtain by her 18th birthday, given the prevailing pattern of enrollment rates across grades, and assuming she starts preschool at age 4. The best possible outcome occurs when children stay in school for 14 years, through age 18. High enrollment rates throughout the school system bring many rich countries close to the 14-year benchmark. But in the poorest countries, children can expect to complete only half of that.

The World Bank Group and its partners are developing a comprehensive new database of international student achievement test scores covering around 160 economies to benchmark what children learn. The database harmonizes results from international and regional testing programs so they are comparable. For the first time, learning is measurable in nearly all countries using the same yardstick.

The differences in learning are dramatic. Country-level average test scores range from around 600 in the best-performing countries to around 300 in the worst-performing. To put these numbers in perspective, a score of roughly 400 corresponds to a benchmark of minimum proficiency set by the Programme for International Student Assessment (PISA), the largest international testing program. Less than half of students in developing countries meet this standard, compared with 86 percent in advanced economies. In Singapore, 98 percent of students reach the international benchmark for basic proficiency in secondary school; in South Africa, only 26 percent of students meet that standard. Essentially, then, all of Singapore's secondary school students are prepared for a postsecondary education and the world of work, while almost three-quarters of South Africa's young people are not.

For health, there is no single directly measured and widely available indicator comparable to years of school as a measure of educational attainment. In the absence of such a measure, two proxies for the overall health environment make up this component of the index: adult survival rates and the rate of stunting for children under age 5. Adult survival rates are used as a proxy for the range of nonfatal health outcomes that a child born in a given year is likely to experience as an adult if current conditions prevail into the future. Stunting measures the share of children who are unusually small for their age. It is broadly accepted as a proxy for the prenatal, infant, and early childhood health environment, and it summarizes the risks to good health that children are likely to experience in their early years—with important consequences for health and well-being in adulthood.

The education and health components of human capital just described have an intrinsic value that is undeniably important—but also undeniably difficult to quantify. It is therefore challenging to combine the components into a single index that meaningfully reflects their contributions to human capital. Many existing indexes of human capital and human development resort to arbitrary aggregation of their components. By contrast, the components of the human capital index are aggregated by first transforming them into measures of their respective contributions to worker productivity relative to a benchmark corresponding to a complete education and full health. This approach follows the development accounting literature.[26] The size of the contributions of education and health to worker productivity is anchored in the extensive microeconometric literature on estimating returns to education and health.

Because the human capital index is measured in terms of the productivity of the next generation of workers relative to the benchmark of complete education and full health, the units of the index have a natural interpretation: a value of x for a country means that the productivity as a future worker of a child born in a given year in that country is only a fraction x of what it could be under the benchmark (table 3.1). This future productivity is divisible into the contributions of the three components of the index, each of

TABLE 3.1 Measuring the productivity as a future worker of a child born in 2018

Maximum productivity = 1

	Component	A country in the		
		25th percentile	50th percentile	75th percentile
		for component X has a value of . . .		
	Component 1: survival			
1	Probability of survival to age 5	0.95	0.98	0.99
A	*Contribution to productivity*	*0.95*	*0.98*	*0.99*
	Component 2: school			
	Expected years of school	9.5	11.8	13.1
	Test score (out of approx. 600)	375	424	503
2	Quality-adjusted years of school	5.7	8.0	10.5
B	*Contribution to productivity*	*0.51*	*0.62*	*0.76*
	Component 3: health			
3	Fraction of children not stunted	0.68	0.78	0.89
4	Adult survival rate	0.79	0.86	0.91
C	*Contribution to productivity*[a]	*0.88*	*0.92*	*0.95*
	Overall human capital index[b]	**0.43**	**0.56**	**0.72**

Source: WDR 2019 team.

Note: "Contribution to productivity" measures how much each component of the index, as well as the overall index, contributes to the expected future productivity as a worker of a child born in 2018 relative to the benchmark of a complete education and full health. A value of x means that productivity is only a fraction x of what it would be under the benchmark of a complete education and full health. Estimates of productivity contributions are anchored in microeconometric evidence on the returns to education and health. "Quality-adjusted years of school" equals the country's test score relative to the global best test score multiplied by the country's expected years of school.

a. C is calculated as the geometric average of the contributions of numbers 3 and 4 to productivity.
b. A × B × C.

which is also expressed in terms of productivity relative to the benchmark. The three components are then multiplied to arrive at the overall index.

Differences in human capital have large implications for the productivity of the next generation of workers. In a country at around the 25th percentile of the distribution of each of the components, a child born in 2018 will be only 43 percent as productive as that child would be under the benchmark of complete education and full health.

The index, because of its units, can be connected in a straightforward fashion to scenarios for future per capita income and growth. Imagine a status quo scenario in which the expected years of quality-adjusted school and level of health, as measured in the index for a given year, persist into the future. Over time, new entrants to the workforce with status quo education and health replace current members of the workforce, until eventually the entire workforce of the future has the expected years of quality-adjusted school and level of health captured in the current human capital index. It is possible to then compare this scenario with one in which the entire future workforce benefits from a complete education and enjoys full health.

In the long run, per capita GDP in this scenario is higher than in the status quo scenario through two channels: the direct effects of higher worker productivity and the indirect effects that reflect the greater investments in physical capital that are induced by having more productive workers. Combining these effects, a country with an index score of x will in the long run have per capita GDP in the status quo scenario that is only a fraction x of what it could be with a complete education and full health. For example, a country with an index of $x = 0.5$ would in the long run have per capita incomes twice as high as the status quo if its citizens enjoyed a complete education and full health. What this means in terms of average annual growth rates depends on the time period. If 50 years—or about two generations—are required for these scenarios to materialize, then a doubling of future per capita income relative to the status quo corresponds to roughly 1.4 percentage points of additional growth per year.

The index measures the amount of human capital that the average child born in 2018 expects to achieve (figure 3.3). However, averages hide a great deal of variation. Most of the components of the index can be disaggregated by gender for most countries so that differences in the prospects of boys versus girls can be observed. Although it is not possible to do so systematically for a large set of countries, in individual countries in which the data are richer, differences in the components of the index across regions and socioeconomic groups can also be illustrated.

The human capital index presented in table 3.2 is the first edition. Like all cross-country benchmarking exercises, it has limitations, with scope for improvement and expansion in subsequent versions. Components of the index such as stunting and test scores are measured only infrequently in some countries and not at all in others. Data on test scores are retrieved from international testing programs in which the age of the test takers and the subjects covered vary. Test scores may not accurately reflect the quality of the entire education system of a country to the extent that test takers are not representative of the population of all students. Reliable measures of the quality of tertiary education do not yet exist, despite the importance of higher education for human capital in a rapidly changing world. Data on enrollment rates, needed to estimate expected school years, often have many gaps and are reported with significant lags. Sociobehavioral skills are not explicitly captured. Adult survival rates are imprecisely estimated in countries where vital registries are incomplete or nonexistent.

One objective of the human capital index is to call attention to these data shortcomings and to galvanize action to remedy them. Improving data takes time. In the interim, and recognizing these limitations, country scores on the index should be interpreted with caution. While providing estimates of how current education and health shape the productivity of future workers, the index is not a finely graduated measurement of small differences between countries. Because it captures outcomes, it is not a checklist of policy actions. The type and scale of interventions required to build human capital are not the same from country to country.

Although there has been significant improvement in the availability of data on education and health outcomes, there is still a long way to go. For example, advanced cognitive and sociobehavioral skills, which are not

FIGURE 3.3 **The human capital index, 2018**

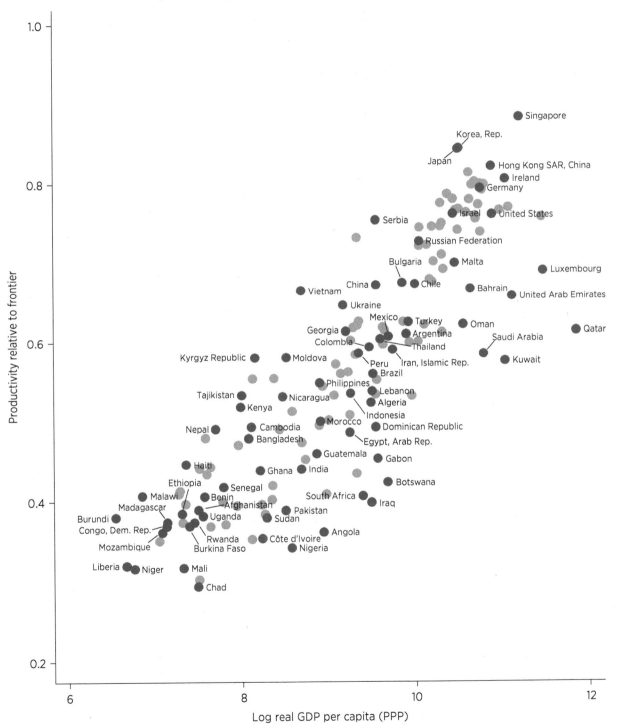

Source: WDR 2019 team.

Note: The human capital index ranges between 0 and 1. The index is measured in terms of the productivity of the next generation of workers relative to the benchmark of complete education and full health. An economy in which the average worker achieves both full health and full education potential will score a value of 1 on the index. GDP = gross domestic product; PPP = purchasing power parity.

TABLE 3.2 The human capital index (HCI), 2018

Rank	Economy	HCI score	Rank	Economy	HCI score	Rank	Economy	HCI score
157	Chad	0.29	104	Egypt, Arab Rep.	0.49	51	Mongolia	0.63
156	South Sudan	0.30	103	Honduras	0.49	50	Ukraine	0.65
155	Niger	0.32	102	Nepal	0.49	49	United Arab Emirates	0.66
154	Mali	0.32	101	Dominican Republic	0.49	48	Vietnam	0.67
153	Liberia	0.32	100	Cambodia	0.49	47	Bahrain	0.67
152	Nigeria	0.34	99	Guyana	0.49	46	China	0.67
151	Sierra Leone	0.35	98	Morocco	0.50	45	Chile	0.67
150	Mauritania	0.35	97	El Salvador	0.50	44	Bulgaria	0.68
149	Côte d'Ivoire	0.35	96	Tunisia	0.51	43	Seychelles	0.68
148	Mozambique	0.36	95	Tonga	0.51	42	Greece	0.68
147	Angola	0.36	94	Kenya	0.52	41	Luxembourg	0.69
146	Congo, Dem. Rep.	0.37	93	Algeria	0.52	40	Slovak Republic	0.69
145	Yemen, Rep.	0.37	92	Nicaragua	0.53	39	Malta	0.70
144	Burkina Faso	0.37	91	Panama	0.53	38	Hungary	0.70
143	Lesotho	0.37	90	Paraguay	0.53	37	Lithuania	0.71
142	Rwanda	0.37	89	Tajikistan	0.53	36	Croatia	0.72
141	Guinea	0.37	88	Macedonia, FYR	0.53	35	Latvia	0.72
140	Madagascar	0.37	87	Indonesia	0.53	34	Russian Federation	0.73
139	Sudan	0.38	86	Lebanon	0.54	33	Iceland	0.74
138	Burundi	0.38	85	Jamaica	0.54	32	Spain	0.74
137	Uganda	0.38	84	Philippines	0.55	31	Kazakhstan	0.75
136	Papua New Guinea	0.38	83	Tuvalu	0.55	30	Poland	0.75
135	Ethiopia	0.38	82	West Bank and Gaza	0.55	29	Estonia	0.75
134	Pakistan	0.39	81	Brazil	0.56	28	Cyprus	0.75
133	Afghanistan	0.39	80	Kosovo	0.56	27	Serbia	0.76
132	Cameroon	0.39	79	Jordan	0.56	26	Belgium	0.76
131	Zambia	0.40	78	Armenia	0.57	25	Macao SAR, China	0.76
130	Gambia, The	0.40	77	Kuwait	0.58	24	United States	0.76
129	Iraq	0.40	76	Kyrgyz Republic	0.58	23	Israel	0.76
128	Tanzania	0.40	75	Moldova	0.58	22	France	0.76
127	Benin	0.41	74	Sri Lanka	0.58	21	New Zealand	0.77
126	South Africa	0.41	73	Saudi Arabia	0.58	20	Switzerland	0.77
125	Malawi	0.41	72	Peru	0.59	19	Italy	0.77
124	eSwatini	0.41	71	Iran, Islamic Rep.	0.59	18	Norway	0.77
123	Comoros	0.41	70	Colombia	0.59	17	Denmark	0.77
122	Togo	0.41	69	Azerbaijan	0.60	16	Portugal	0.78
121	Senegal	0.42	68	Uruguay	0.60	15	United Kingdom	0.78
120	Congo, Rep.	0.42	67	Romania	0.60	14	Czech Republic	0.78
119	Botswana	0.42	66	Ecuador	0.60	13	Slovenia	0.79
118	Timor-Leste	0.43	65	Thailand	0.60	12	Austria	0.79
117	Namibia	0.43	64	Mexico	0.61	11	Germany	0.79
116	Ghana	0.44	63	Argentina	0.61	10	Canada	0.80
115	India	0.44	62	Trinidad and Tobago	0.61	9	Netherlands	0.80
114	Zimbabwe	0.44	61	Georgia	0.61	8	Sweden	0.80
113	Solomon Islands	0.44	60	Qatar	0.61	7	Australia	0.80
112	Haiti	0.45	59	Montenegro	0.62	6	Ireland	0.81
111	Lao PDR	0.45	58	Bosnia and Herzegovina	0.62	5	Finland	0.81
110	Gabon	0.45	57	Costa Rica	0.62	4	Hong Kong SAR, China	0.82
109	Guatemala	0.46	56	Albania	0.62	3	Japan	0.84
108	Vanuatu	0.47	55	Malaysia	0.62	2	Korea, Rep.	0.84
107	Myanmar	0.47	54	Oman	0.62	1	Singapore	0.88
106	Bangladesh	0.48	53	Turkey	0.63			
105	Kiribati	0.48	52	Mauritius	0.63			

Source: WDR 2019 team.

Note: The human capital index ranges between 0 and 1. The index is measured in terms of the productivity of the next generation of workers relative to the benchmark of complete education and full health. An economy in which the average worker achieves both full health and full education potential will score a value of 1 on the index.

incorporated in the index, are important contributors to individual productivity. And comparable data are lacking on early childhood development, which is a significant foundation for the quality of the future labor force.

Yet another task is measuring the intermediate factors that affect these outcomes. Although citizens of low- and middle-income countries face similar constraints in the accumulation of human capital, the relevance of these constraints is often context-specific. Understanding which constraints matter the most is essential to setting priorities across policy areas.

A first step is improving the quality of basic administrative data in education and health. Only one in six governments publish annual education monitoring reports. Just 100 countries or so report reasonably complete and up-to-date data on net enrollment rates at different levels of education to the UNESCO Institute for Statistics—the body tasked with compiling this data internationally. Monitoring of even the most basic health information—births and deaths—is insufficient in low- and middle-income countries (figure 3.4). The pace of improvement in these systems has been slow. Worldwide, between 2000 and 2012 the percentage of registered deaths changed from only 36 percent to 38 percent. The percentage of children under 5 whose births were registered only increased from 58 percent to 65 percent.[27] High-quality basic administrative data are essential for governments to understand their needs and to plan the allocation of public services.

Increasing the number of countries in which the learning achievements of children are measured—both those in and out of school—would allow much better tracking of countries' performance for school access and learning. This should include making data on learning fully representative of all children rather than the selection—often from higher-income families—of those who stay in school. The *Annual Status of Education Report* is a rare example of a survey that provides an annual assessment of the learning levels of children—in this case from India's rural households—of those who are also out of school.

Initiatives that create comparable measures of learning across countries would be a remedy.

FIGURE 3.4 Records of births and deaths remain inadequate

Civil registrations of births and deaths by country income group, 2018

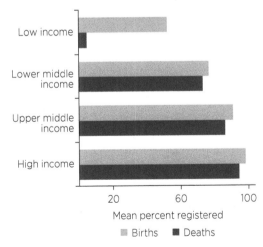

Mean percent registered

■ Births   ■ Deaths

Source: WDR 2019 team.

Note: Figure shows estimates of birth and death registration coverage based on available data for 180 and 120 countries, respectively. Birth registration data are based on the *United Nations Demographic Yearbook.* For countries with incomplete civil registration systems, birth registration is estimated from mothers' self-reporting of their children's birth registration status, as collected in household surveys. Death registration data are based on estimates by the World Health Organization.

They would seek to bring together stakeholders to agree on a set of common questions to include in learning assessments, thereby allowing results to be harmonized across tests. In the short term, the existing data platforms—national household surveys, Demographic and Health Surveys, Living Standards Measurement Study, and Service Delivery Indicators—could be used to increase the availability of data on human capital outcomes in a cost-effective way.

Similar efforts are under way in health. To improve the coordination of health data collection, the Health Data Collaborative was launched in 2015 by a group of international agencies, bilateral and multilateral donors, foundations, and governments. New technologies such as the global positioning system and mobile phones are driving down costs of data collection. The Primary Health Care Performance Initiative, launched by the Bill & Melinda Gates Foundation, World Bank Group, and World Health Organization in 2015, provides an international benchmarking of primary care quality.

A second step is to better understand the many dimensions of socio-behavioral and other skills, health, and the correlation between the two. Sociobehavioral skills are multidimensional. Initiatives such as the World Bank's Skills Towards Employability and Productivity surveys and the Organisation for Economic Co-operation and Development's Programme for the International Assessment of Adult Competencies surveys have sought to measure these skills on a large scale among working-age individuals. There has not been a similar attempt among school-age children, even though there is evidence that abilities such as grit and self-regulation matter for learning. Interventions that have reduced iron deficiency anemia have been found to improve student learning outcomes, but the correlation between health status and student test scores has not yet been quantified. The introduction of health modules in school surveys would be an important first step. Relatively low-cost assessments, such as those of student vision acuity and anthropometric status, can go a long way toward improving understanding of the relationship between learning and health.

Vietnam's experience illustrates the potential benefits of mapping pathways of change. The country's schoolchildren scored in the top quarter of the mostly middle- and high-income countries that participated in the 2012 and 2015 PISA. This performance is remarkable in view of Vietnam's level of per capita income. Understanding this success could provide important lessons for how to ensure that schooling achieves learning.

As the nature of work changes, human capital becomes more important. Yet significant gaps in human capital persist across the world. These gaps—manifested in low education and health outcomes—hurt the future productivity of workers and future competitiveness of economies. To address this issue, governments must seek remedies. However, because of the long time needed for human capital investments to yield economic returns, the political incentives for human capital investments are often missing. The human capital project aims to create not just these incentives, but also the policy guidance for more and better investments in human capital.

Notes

1. Fernald and Hidrobo (2011).
2. Smith ([1776] 1937, book 2, chap.1), as reported in Goldin (2016).
3. Chetty, Friedman, and Rockoff (2014).
4. Psacharopoulos and Patrinos (2018).
5. Dillon, Friedman, and Serneels (2014).
6. Ahuja et al. (2015).
7. Belot and James (2011).
8. Sandjaja et al. (2013).
9. Dillon et al. (2017).
10. Flabbi and Gatti (2018).
11. Ahuja et al. (2015).
12. Andrabi, Das, and Khwaja (2012).
13. Hsieh and Klenow (2010).
14. Cavaillé and Marshall (2017).
15. Knack and Keefer (1997).
16. Rosas and Sabarwal (2016).
17. Blanford et al. (2012).
18. Ferré and Sharif (2014).
19. Baird, McIntosh, and Özler (2016).
20. Jensen (2010).
21. Giné, Karlan, and Zinman (2010).
22. Heckman et al. (2010).
23. Coffey, Geruso, and Spears (2018).
24. Gertler et al. (2014).
25. Kraay (2018).
26. Caselli (2005); Weil (2007).
27. Mikkelsen et al. (2015).

References

Ahuja, Amrita, Sarah Baird, Joan Hamory Hicks, Michael R. Kremer, Edward Miguel, and Shawn Powers. 2015. "When Should Governments Subsidize Health? The Case of Mass Deworming." *World Bank Economic Review* 29 (supplement 1): S9–S24.

Andrabi, Tahir, Jishnu Das, and Asim Ijaz Khwaja. 2012. "What Did You Do All Day? Maternal Education and Child Outcomes." *Journal of Human Resources* 47 (4): 873–912.

Baird, Sarah Jane, Craig T. McIntosh, and Berk Özler. 2016. "When the Money Runs Out: Do Cash Transfers Have Sustained Effects on Human Capital Accumulation?" Policy Research Working Paper 7901, World Bank, Washington, DC.

Belot, Michèle, and Jonathan James. 2011. "Healthy School Meals and Educational Outcomes." *Journal of Health Economics* 30 (3): 489–504.

Blanford, Justine I., Supriya Kumar, Wei Luo, and Alan M. MacEachren. 2012. "It's a Long, Long Walk: Accessibility to Hospitals, Maternity, and Integrated Health Centers in Niger." *International Journal of Health Geographics* 11 (24): 1–15.

Caselli, Francesco. 2005. "Accounting for Cross-Country Income Differences." In *Handbook of Economic Growth*, vol. 1A, edited by Philippe Aghion and Steven N. Darlauf, 679–741. Amsterdam: Elsevier.

Cavaillé, Charlotte, and John Marshall. 2017. "Education and Anti-immigration Attitudes: Evidence from Compulsory Schooling Reforms across Western Europe." Working paper, Georgetown University, Washington, DC, December.

Chetty, Raj, John N. Friedman, and Jonah E. Rockoff. 2014. "Measuring the Impacts of Teachers II: Teacher Value-Added and Student Outcomes in Adulthood." *American Economic Review* 104 (9): 2633–79.

Coffey, Diane, Michael Geruso, and Dean Spears. 2018. "Sanitation, Disease Externalities, and Anaemia: Evidence from Nepal." *Economic Journal* 128 (611): 1395–1432.

Dillon, Andrew, Jed Friedman, and Pieter Serneels. 2014. "Health Information, Treatment, and Worker Productivity: Experimental Evidence from Malaria Testing and Treatment among Nigerian Sugarcane Cutters." Policy Research Working Paper 7120, World Bank, Washington, DC.

Dillon, Moira R., Harini Kannan, Joshua T. Dean, Elizabeth S. Spelke, and Esther Duflo. 2017. "Cognitive Science in the Field: A Preschool Intervention Durably Enhances Intuitive but Not Formal Mathematics." *Science* 357 (6346): 47–55.

Fernald, Lia C. H., and Melissa Hidrobo. 2011. "Effect of Ecuador's Cash Transfer Program (Bono de Desarrollo Humano) on Child Development in Infants and Toddlers: A Randomized Effectiveness Trial." *Social Science and Medicine* 72 (9): 1437–46.

Ferré, Céline, and Iffath Sharif. 2014. "Can Conditional Cash Transfers Improve Education and Nutrition Outcomes for Poor Children in Bangladesh? Evidence from a Pilot Project." Policy Research Working Paper 7077, World Bank, Washington, DC.

Flabbi, Luca, and Roberta Gatti. 2018. "A Primer on Human Capital." Policy Research Working Paper 8309, World Bank, Washington, DC.

Gertler, Paul J., James J. Heckman, Rodrigo Pinto, Arianna Zanolini, Christel Vermeersch, Susan P. Walker, Susan M. Chang, and Sally M. Grantham-McGregor. 2014. "Labor Market Returns to an Early Childhood Stimulation Intervention in Jamaica." *Science* 344 (6187): 998–1001.

Giné, Xavier, Dean Karlan, and Jonathan Zinman. 2010. "Put Your Money Where Your Butt Is: A Commitment Contract for Smoking Cessation." *American Economic Journal: Applied Economics* 2 (4): 213–35.

Goldin, Claudia. 2016. "Human Capital." In *Handbook of Cliometrics*, edited by Claude Diebolt and Michael John Haupert, 55–86. Berlin: Springer.

Heckman, James J., Seong Hyeok Moon, Rodrigo Pinto, Peter A. Savelyev, and Adam Yavitz. 2010. "The Rate of Return to the HighScope Perry Preschool Program." *Journal of Public Economics* 94 (1–2): 114–28.

Hsieh, Chang-Tai, and Peter J. Klenow. 2010. "Development Accounting." *American Economic Journal: Macroeconomics* 2 (1): 207–23.

Jensen, Robert 2010. "The (Perceived) Returns to Education and the Demand for Schooling." *Quarterly Journal of Economics* 125 (2): 515–48. https://econpapers.repec.org/article/oupqjecon/.

Knack, Stephen, and Philip Keefer. 1997. "Does Social Capital Have an Economic Payoff? A Cross-Country Investigation." *Quarterly Journal of Economics* 112 (4): 1251–88.

Kraay, Aart. 2018. "Methodology for a World Bank Human Capital Index." Policy Research Working Paper 8593, World Bank, Washington, DC.

Mikkelsen, Lene, David E. Phillips, Carla AbouZahr, Philip W. Setel, Don de Savigny, Rafael Lozano, and Alan D. Lopez. 2015. "A Global Assessment of Civil Registration and Vital Statistics Systems: Monitoring Data Quality and Progress." *Lancet* 386 (10001): 1395–1406.

Patrinos, Harry Anthony, and Noam Angrist. 2018. "A Global Dataset on Education Quality: A Review and an Update (1965–2018)." Policy Research Working Paper 8592, World Bank, Washington, DC.

Psacharopoulos, George, and Harry Anthony Patrinos. 2018. "Returns to Investment in Education: A Decennial Review of the Global Literature." Policy Research Working Paper 8402, World Bank, Washington, DC.

Rosas, Nina, and Shwetlena Sabarwal. 2016. "Can You Work It? Evidence on the Productive Potential of Public Works from a Youth Employment Program in Sierra Leone." Policy Research Working Paper 7580, World Bank, Washington, DC.

Sandjaja, Bee Koon Poh, Nipa Rojroonwasinkul, Bao Khanh Le Nyugen, Basuki Budiman, Lai Oon Ng, Kusol Soonthorndhada, Hoang Thi Xuyen, Paul Deurenberg, et al. 2013. "Relationship between Anthropometric Indicators and Cognitive Performance in Southeast Asian School-Aged Children." *British Journal of Nutrition* 110 (supplement 3): S57–S64.

Smith, Adam. [1776] 1937. *An Inquiry into the Nature and Causes of the Wealth of Nations*, book 2. Modern Library Series Reprint. New York: Random House.

Weil, David N. 2007. "Accounting for the Effect of Health on Economic Growth." *Quarterly Journal of Economics* 122 (3): 1265–1306.

CHAPTER 4

Lifelong learning

Nelson Mandela, the first president of postapartheid South Africa, once said, "Education is the great engine of personal development. It is through education that the daughter of a peasant can become a doctor, that the son of a mineworker can become the head of the mine, that the child of a farmworker can become the president of a great nation. It is what we make out of what we have, not what we are given, that separates one person from another."

Automation is reshaping work and the skills demanded for work. The demand for advanced cognitive skills[1] and sociobehavioral skills[2] is increasing, whereas the demand for narrow job-specific skills is waning.[3] Meanwhile, the skills associated with "adaptability" are increasingly in demand. This combination of specific cognitive skills (critical thinking and problem-solving) and sociobehavioral skills (creativity and curiosity) is transferable across jobs.

How well countries cope with the demand for changing job skills depends on how quickly the supply of skills shifts. Education systems, however, tend to resist change. A significant part of the readjustment in the supply of skills is happening outside of compulsory education and formal jobs. Early childhood learning, tertiary education, and adult learning sought outside the workplace are increasingly important in meeting the skills that will be sought by future labor markets. This chapter shows how.

Automation—and the adoption of technology more generally—makes some jobs obsolete. The demand for skills linked to home appliance repair, for example, is shrinking quickly because technology is driving down the price of appliances and improving reliability. At the same time, innovation is creating new types of jobs. In fact, a large share of children entering primary school in 2018 will work in occupations that do not yet exist. Even in low- and middle-income countries, many people are employed in jobs that did not exist three decades ago. India has nearly 4 million app developers; Uganda has over 400,000 internationally certified organic farmers; and China has 100,000 data labelers.

Meanwhile, many current jobs are being retooled into new forms, resulting in new and sometimes unexpected skill combinations. In 2018 a marketing professional might well be asked to write algorithms. A physics graduate may land a job as a quantitative trader in the finance industry. Workers who bring emerging skills into relevant technical fields of expertise—such as teachers who are good at web design and actuaries who are proficient in big data analytics—are likely to be in high demand.

Which skills are in less demand in 2018? Evidence from developed countries points to job polarization—the expansion of high- and low-skill jobs coupled with the decline of middle-skill jobs. The demand for workers who can undertake nonroutine cognitive tasks, such as high-skilled research, is increasing. So is the relative demand for workers able to handle nonroutine tasks that cannot be automated easily, such as food preparation. Conversely, the demand for workers for procedural routine tasks, which are often performed in middle-skill jobs such as data entry, is declining because of automation.

FIGURE 4.1 In many developing countries, the share of employment in high-skill occupations has increased

Annual average change in employment share, by occupation skill level, circa 2000–circa 2015

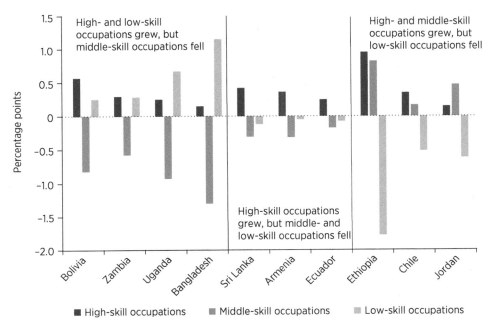

Source: WDR 2019 team, based on World Bank's International Income Distribution Data Set.

Note: High-skill occupations: managers, professionals, technicians, and associate professionals. Middle-skill occupations: clerical support workers; sales and services workers; craft and related trades workers; skilled agricultural, forestry, and fishery workers; plant and machine operators and assemblers. Low-skill occupations: elementary occupations such as cleaners and helpers; laborers in agriculture, forestry, and fisheries; laborers in mining, construction, manufacturing, and transport; food preparation assistants; street and related sales and services workers.

Is the same pattern beginning to emerge in low- and middle-income countries? Not quite. In many developing countries, the demand for high-skill workers is increasing (figure 4.1). The share of workers in high-skill occupations increased by 8 percentage points or more in Bolivia, Ethiopia, and South Africa from 2000 to 2014. But the change in demand for low- and middle-skill jobs is more heterogeneous across countries. In Jordan, the share of employment in middle-skill jobs increased by 7.5 percentage points between 2000 and 2016. In Bangladesh, this share fell by almost 20 percentage points during the same period.[4]

This change in the demand for workers for low- and middle-skill jobs in developing countries is not surprising. What happens at this end of the skills spectrum is likely to be driven by the competing forces of automation and globalization. The rate of technology adoption tends to vary considerably across developing countries. In Europe and Central Asia, 26 percent of the population had fixed broadband subscriptions in 2016, compared with just 2 percent in South Asia. Globalization is bringing the low- and medium-skill jobs of developed countries to some—but not all—developing countries. Depending on the relative speed of these forces, some developing countries are seeing an increase in middle-skill jobs; others are seeing a decline.

Creating a skilled workforce for the future of work rests on the growing demand for advanced cognitive skills, sociobehavioral skills, and adaptability. Evidence across low- to high-income countries suggests that in recent decades jobs are being defined by more cognitive, analytical tasks. In Bolivia and Kenya, more than 40 percent of workers using computers perform complex tasks that require advanced programming. Indeed, the demand is growing for transferable higher-order cognitive skills such as logic, critical thinking, complex problem-solving, and reasoning. In all regions of the world, these skills are consistently ranked among those most valued by employers. Analysis of the job markets in Denmark, France, Germany, the Slovak Republic, South Africa, Spain, and Switzerland reveals that a one standard deviation increase in complex problem-solving skills is associated with a 10–20 percent higher wage.[5] In Armenia and Georgia, the ability to solve problems and learn new skills yields a wage premium of nearly 20 percent.[6]

The demand for sociobehavioral skills is also increasing in developing countries. In Latin America and the Caribbean, the adoption of digital technology has placed more importance on general cognitive skills and raised the demand for workers with interpersonal skills. In Cambodia, El Salvador, Honduras, the Lao People's Democratic Republic, Malaysia, the Philippines, and Vietnam, more than half of firms report shortages of workers with specific sociobehavioral skills, such as commitment to work.[7]

Technological change makes it harder to anticipate which job-specific skills will thrive and which will become obsolete in the near future. In the past, shifts in skill requirements prompted by technological progress took centuries to manifest themselves (figure 4.2). In the digital era, advances in technology call for new skills seemingly overnight.

The ability to adapt quickly to changes is increasingly valued by the labor market. The sought-after trait is adaptability—the ability to respond to unexpected circumstances and to unlearn and relearn quickly. This trait requires a combination of certain cognitive skills (critical thinking, problem-solving) and sociobehavioral skills (curiosity, creativity). A study of technical and vocational students in Nigeria showed that the sociobehavioral skill of self-efficacy was positively and significantly predictive of career adaptability.[8]

Strong skill foundations are important for developing advanced cognitive skills, sociobehavioral skills, and skills predictive of adaptability. For most children, these skill foundations are formed through primary and secondary education. Yet, according to the *World Development Report 2018*, the acquisition of foundational skills that one would expect to happen in schools is not occurring in many low- and middle-income countries.[9]

Important skills readjustments are happening increasingly outside of compulsory education and formal jobs. Skills development for the changing nature of work is a matter of lifelong learning. This kind of learning is especially germane to skills readjustment amid demographic change—be it the aging populations of East Asia and Eastern Europe or the large youth populations of Sub-Saharan Africa and South Asia.

FIGURE 4.2 The rate of technology diffusion is increasing

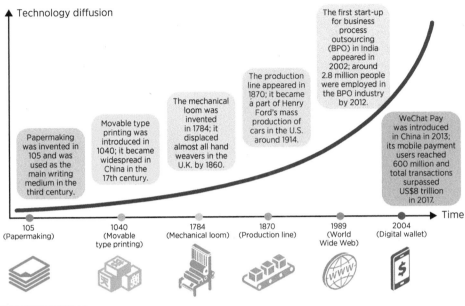

Source: WDR 2019 team.

Learning in early childhood

In France, the mandatory school starting age will soon be lowered from 6 to 3 years. According to President Emmanuel Macron, this reform is intended to boost equality, thereby improving the ability of children from disadvantaged backgrounds to remain competitive in the education system.

The most effective way to acquire the skills demanded by the changing nature of work is to start early. Early investments in nutrition, health, social protection, and education lay strong foundations for the future acquisition of cognitive and sociobehavioral skills. They also make future skills acquisition more resilient to uncertainty. Early childhood investments are an important way to improve equality of opportunity. Currently, these investments are underprovided, especially for poor, disadvantaged children, who would benefit from them the most. Prioritizing these investments could pay off significantly for economies, as long as both access and quality are highlighted.

The architecture of the brain forms from the prenatal period to age 5, and so this is an important stage for developing cognitive and sociobehavioral skills. During this period, the brain's ability to learn from experience is at its highest level (figure 4.3). Experiences and learning during this period directly affect achievement in adulthood. If this window is missed, building skills becomes harder.

Quality early childhood development programs enable children to learn. Investments in nutrition, health, and stimulation in the first thousand days of life build stronger brains. The engagement of parents and caregivers during this phase also matters for the development of children's language

FIGURE 4.3 **The brain's ability to learn from experience decreases with age**

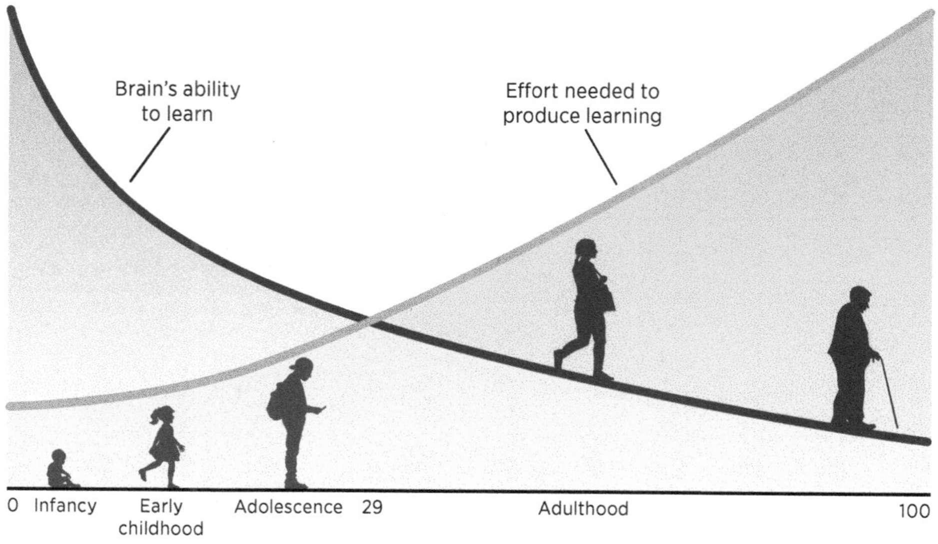

Brain's ability
to learn

Effort needed to
produce learning

| 0 Infancy | Early childhood | Adolescence | 29 | Adulthood | 100 |

Source: WDR 2019 team.

skills, motor and self-regulation skills, as well as social behavior. In Colombia, exposure to psychological stimulations through home visits with play demonstrations significantly improved the cognitive development of children ages 12–24 months.[10] In Pakistan, the efforts of the Lady Health Worker Programme, which provides health services in rural areas, led to children under 3 years old being more likely to be fully immunized in 2008 than in 2000 by 15 percentage points.[11] The program has generated sustained positive effects on children's cognitive abilities and pro-social behaviors by providing nutrition supplementation and encouraging mothers to engage in responsive play with children up to age 2.

From the age of 3, socialization and more formal early learning become important to prepare children to succeed in primary school. A quality preschool strengthens children's executive functions (such as working memory, flexible thinking, self-control), launching them on higher learning trajectories. In Bangladesh, rural children who attended preschool performed better in early-grade speaking, writing, and mathematics, compared with those who did not.[12] A preschool reform in rural Mozambique had positive effects on sociobehavioral development—participating children were better at interacting with others, following directions, and regulating their emotions under stress.[13] However, to achieve these results preschools have to meet quality thresholds. In some cases, a low-quality preschool is worse for child development than no preschool at all.[14]

Poor-quality early childhood development programs are associated with disappointing results in children's language development, cognitive skills, and sociability. A study of preschools in a Nairobi slum in Kenya revealed that, despite high participation rates, the curriculum and pedagogical approach were not age-appropriate. In the program, 3- to 6-year-olds had to follow academic-oriented instruction and even sit for exams.[15] In Peru, although the national Wawa Wasi program has provided safe community-based daycare and a nutritious diet for children from ages 4 to 6 in impoverished areas, it has failed to improve children's language or motor development skills because of insufficiently trained caregivers.

Early childhood investments efficiently produce skills that are relevant to a child's future. Learning is cumulative—skills acquired at an earlier stage facilitate skills formation in subsequent stages. The returns to early investments are the highest of those made over the life span, and the advantages conferred by these investments grow over time. An additional dollar invested in quality early childhood programs yields a return of US$6–$17.[16]

Early childhood development programs improve parents' labor force participation. Many women do not work because of time-consuming child-rearing responsibilities. In the United Kingdom, half of stay-at-home mothers would prefer to go back to work if they had access to high-quality, affordable child care services. Early investments in such services would alleviate this constraint. In Argentina, a large-scale construction program of pre-primary school facilities in the 1990s positively affected maternal employment. In Spain, during the same period, maternal employment increased by 10 percent because of the availability of full-time public care for 3-year-olds.[17]

Early childhood investments also increase equity. For children exposed to poverty and other adverse conditions, quality early childhood programs increase adult competence, reduce violent behavior and social inhibition, and foster growth in the subsequent generation. In Guatemala, an early childhood development nutrition program for poor families significantly increased the wages for these children in adulthood.[18] In Jamaica, early stimulation for infants and toddlers increased their future earnings by 25 percent—equivalent to that of adults who grew up in wealthier households.[19]

Despite their efficiency in producing important skills, early childhood investments are underprovided. Some 250 million children under age 5 are at risk of not reaching their developmental potential in low- and middle-income countries because of stunting or extreme poverty. Worldwide, more than 87 million children under age 7 have spent their entire lives in conflict-affected areas. They suffer from extreme trauma and toxic stress, which impair their brain development and skill enhancement. Only half of all 3- to 6-year-olds have access to preprimary education globally—in low-income countries this share is one-fifth. In 2012 North America and Western Europe spent 8.8 percent of their education budgets on preprimary education; in Sub-Saharan Africa the share allocated was only 0.3 percent.

Children from poor families are the least likely to attend early childhood development programs (figure 4.4). They would also likely benefit the most from such programs. In low- and middle-income countries, approximately 47 percent of the wealthiest families have access to early education programs, but for the poorest families this number is 20 percent.[20] Rural families are especially disadvantaged. Across a sample of 14 low- and middle-income

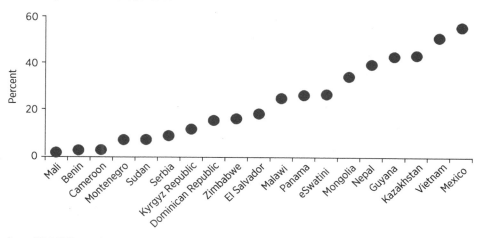

FIGURE 4.4 In many countries, children from disadvantaged backgrounds are the least likely to attend early childhood education programs

Proportion of children ages 3–4 years attending early education programs among the poorest 20 percent of households, selected countries, circa 2014

Source: WDR 2019 team, based on data obtained from UNICEF Multiple Indicator Cluster Survey.

countries, rural dwellers consistently have worse access to early childhood development programs, compared with those living in urban areas.[21]

Effective pathways to early childhood development are available. In some places, community-based playgroups have produced consistently positive low-cost results. In Indonesia, a playgroup program positively affected children's language, sociobehavioral, and cognitive skills. Children from disadvantaged backgrounds benefited more in both the short and long term.[22] In Tonga, organizing playgroups for children up to age 5 significantly improved their early-grade reading skills. The Montessori model, characterized by multiage classrooms, student-chosen learning activities, and minimal instruction, has been shown to be more effective than conventional education in improving children's executive functions. With successful local adaptations, Montessori and other child-centered approaches—including Steiner, Reggio Emilia, and Tools of the Mind—can be found in diverse settings, from Haiti to Kenya.

Research has uncovered several concrete ways to increase take-up of early childhood development investments. Cash transfers that support early childhood development for the poorest children have succeeded in various contexts. Such programs have reduced stunting in the Philippines and Senegal, fostered language development in Ecuador and Mexico, and improved children's sociobehavioral skills in Niger. Integrated approaches that combine health, nutrition, and stimulation investments have also been highly effective. Chile's Crece Contigo program integrates the services provided by the health, education, welfare, and protection services so that a child's first contact with the system occurs while still in the womb during the mother's first prenatal visit.

Tertiary education

The Free University of Tbilisi was established in 2007 through a nonprofit organization. It has already become the top-performing, most sought-after university in Georgia. This success stems from a transparent admissions process (national competitive entry examinations), as well as a competitive state financing program for individual students based on academic performance. Per capita financing increases the efficiency and transparency of university financing, allowing the government to gradually reduce lump-sum payments to universities directly. The university offers a high-quality faculty, flexible course offerings, and discussion-based pedagogy. Each year, the university attracts hundreds of top-tier applicants, and more than 96 percent of its graduates find employment or enroll in further education.

Integrated, technology-driven economies are increasingly valuing tertiary education (defined as any education beyond the high school level, including trade schools and college). The global average private return to tertiary education is 16 percent.[23] But such a return is not high for everyone. It depends on a range of factors that include the quality of the provider, the composition of the student population, and the availability of jobs. Controlling for other factors, students attending a top university in Colombia earn 20 percent more than those who just failed to achieve the score required for university admission.[24] The return also varies dramatically depending on the specialization. In Chile, the return to tertiary education ranges from 4 percent for humanities to 126 percent for engineering and technology.[25] Tertiary enrollment and expenditure also vary considerably by region (figure 4.5).

The changing nature of work makes tertiary education more attractive in three ways. First, technology and integration have increased the demand for higher-order general cognitive skills—such as complex problem-solving, critical thinking, and advanced communication—that are transferable across jobs but cannot be acquired through schooling alone. The rising demand for these skills has enhanced the wage premiums of tertiary graduates, while reducing the demand for less educated workers. Second, tertiary education increases the demand for lifelong learning. Workers are expected to have multiple careers, not just multiple jobs over their lifetime. Tertiary education—with its wide array of course offerings and flexible delivery models such as online learning and open universities—meets this growing demand. Third, tertiary education—especially universities—becomes more attractive in the changing world of work by serving as a platform for innovation.

The relevance of tertiary education systems for the future of work depends on how well they deliver on these three fronts. Increasingly, skills acquisition is a continuum, not a finite, unchangeable path. Flexibility is increased by ensuring that when students open the door to one pathway, the doors to other pathways do not close irrevocably. For example, when first undertaking a tertiary education most students must choose between general education and vocational training. General education such as programs on engineering

FIGURE 4.5 The gross tertiary enrollment ratio and percentage expenditure on tertiary education varied by region in 2016

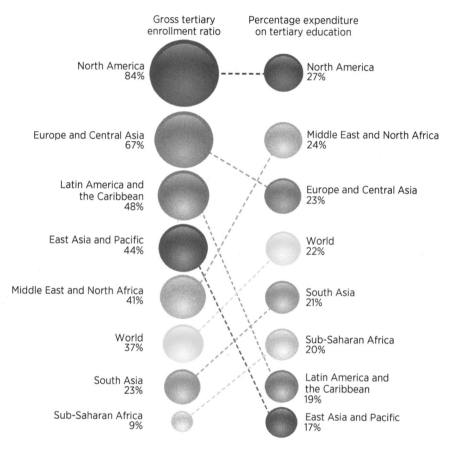

Sources: World Bank's World Development Indicators database. Data on tertiary education expenditure for the Middle East and North Africa are from World Bank (2018a).

Note: "Gross tertiary enrollment ratio" is the ratio of total enrollment, regardless of age, to the population of the age group that officially corresponds to tertiary education. "Percentage expenditure on tertiary education" is the expenditure on tertiary education expressed as a percentage of the total general government expenditure on education.

or economics equip students with the transferable higher-order skills that determine their overall learning readiness or trainability. By contrast, vocational training, such as programs on nursing or airport operations, is directly related to specific occupations. Once this choice is made—especially if it is for vocational training—it is typically difficult and expensive to reverse.

The relative returns to general and vocational education are changing in unpredictable ways, and most economies continue to demand both. Technological progress tends to lower the demand for certain occupation-specific skills, making certain vocational degrees obsolete. It also leads to a higher depreciation of narrow job-specific skills compared with general skills. Yet vocational training continues to be pursued by many students. In 2012 63 percent of Dutch higher education students were attending vocational training.[26] This share was more than 50 percent in Malaysia and 31 percent in Kenya in 2013.[27] Vocational training meets the immediate demand for technical skills, enables faster education-to-work transitions for some, and alleviates pressure on the university system.

Three factors make flexibility between the general and technical tracks imperative for the changing nature of work. First, the combination of general and technical skills is becoming highly valued. Second, even technical jobs seem to be requiring more and more intensive higher-order general skills, implying that this type of skills acquisition should be accessible before and during one's working life. Third, people trained in narrow vocational skills would benefit from opportunities to gain new skills. For example, the Democratic Republic of Congo and Tanzania offer "bridging" arrangements that enable vocational graduates to continue to university.

Close collaboration between industry and vocational education also plays a role. In China, Lenovo is working with tertiary institutes to train vocational students in high-tech areas, such as cloud computing, that feature practice-based curricula, practitioner-led instruction, and professional certification. Filling in information gaps enables students to make choices between and within different paths. Chile is establishing online platforms where students can access information on the employability of individuals with various degrees, wage profiles, and courses to take for certain occupations.

Tertiary systems have not remained impervious to these changing demands—general and vocational tracks often intersect. A wide range of programs offered by universities have a vocational dimension or orientation, including many in science, engineering, and technology. Technology-enabled platforms are making tertiary education more available, especially for those with historically low access. The five largest distance-learning programs are based in lower- or middle-income countries. India is the second-largest consumer of massive open online courses (MOOCs). XuetangX, China's biggest MOOC and blended learning portal, was serving 10 million students in 2018. In Brazil, Veduca launched the world's first open online master's in business administration program in 2013 and was offering over 5,000 courses in 2018. MOOCs are a promising way of delivering flexible, personalized education to a large population. But ensuring quality is a serious challenge. According to a recent study, students who took a course online performed worse than those who followed in-person instructions. Besides content, many MOOCs fail on student engagement or instructor quality.

Tertiary education systems should guarantee a minimum threshold of transferable cognitive skills, which are the best inoculation against job uncertainty. But not all systems are effective at producing these skills. Colombia's universities vary significantly in their ability to impart foundational higher-order skills such as critical thinking, problem-solving, and communication. A study of Chinese undergraduates in engineering and computer science suggests that their cognitive skills did not improve much during the first two years of college.

Incorporating more general education in tertiary programs is one way to increase the acquisition of transferable higher-order cognitive skills. An additional year of general education was added in 2012 to undergraduate programs in Hong Kong SAR, China, focusing on problem-solving, critical thinking, communication, leadership, and lifelong learning skills. For a large majority of students, this change seems to be effectively promoting desirable graduate attributes. Another way is through innovative pedagogy. The Faculty of Architecture and Environmental Design at the College of Science and Technology–University of Rwanda has promoted learning strategies that include open-ended assessment, feedback opportunities, and a progressive curriculum that balances academic challenge with student support. These approaches have improved the critical thinking skills of students.

Tertiary education also builds transferable sociobehavioral skills such as teamwork, resilience, self-confidence, negotiation, and self-expression. In a survey of employers of engineers in India, sociobehavioral skills were ranked at or above technical qualifications and credentials in terms of their significance for the employability of recent graduates. Employer surveys in Bulgaria, Georgia, Kazakhstan, the former Yugoslav Republic of Macedonia, Poland, the Russian Federation, and Ukraine indicate that employers see a lack of sociobehavioral skills as at least as problematic as a lack of technical skills.

Forward-looking universities are finding ways for adult students to acquire sociobehavioral skills. Dutch vocational colleges are providing entrepreneurial courses aimed at improving noncognitive skills such as teamwork and self-confidence. In Tunisia, introducing an entrepreneurship track that combines business training with personal coaching reshaped the behavioral skills of university students. In China, a combination of cooperative learning and role play enhanced self-educational abilities and communication skills among undergraduate students in pharmacology classes. However, better teaching of sociobehavioral skills does require the appropriate curricula and accurate measurement, especially in the context of low-income countries and rural areas.

The role of tertiary education systems as centers of innovation is highly valued as well. Well-known examples of successful university innovation clusters are located in the developed world—in the United States at Stanford University, University of California, Berkeley (Silicon Valley), and Harvard–Massachusetts Institute of Technology (Boston's Route 128); in the United Kingdom at the University of Cambridge–University of Oxford–University College London (the "golden triangle"). Clusters are also emerging in middle-income countries. The University of Malaya in Malaysia has established eight interdisciplinary research clusters over the last decade covering sustainability science and biotechnology. Peking University is building Clinical Medicine Plus X, a research cluster for precision medicine, health big data, and intelligence medicine. As part of the Startup India initiative, seven new research parks have been established on Indian Institute of Technology campuses to promote innovation through incubation and collaboration between universities and private sector firms. In Mexico, the Research and Technology Innovation Park currently houses more than 30 research centers covering research and development in biotechnology, nanotechnology, and robotics. Seven of the centers are led by universities.

Two main factors matter for a healthy innovation ecosystem. First, prioritize the right university for the right sector. The agglomeration effects of universities vary by sector. Second, recognize that a healthy innovation ecosystem requires an enabling environment. Just because successful innovation clusters exist does not mean there is a guaranteed formula for their creation. However, governments are often responsible for creating the enabling environment in which innovation clusters flourish by providing local infrastructure, increasing the expenditure on research and development, connecting universities with high-quality researchers and private sector innovation, and easing rigid labor market regulations.

Adult learning outside the workplace

As the nature of work changes, some workers are caught in the crosshairs of ongoing disruptions in the skills required. As economies rejig to provide the human capital of the next generation, the current working-age population becomes anxious about its job prospects.

One step toward lessening this anxiety is adult learning aimed at supplying workers who are not in school or in jobs with new or updated skills. However, this approach has shown more promise in theory than in practice. Bad design too often gets in the way. Adult learning can be improved in three ways: more systematic diagnoses of the specific constraints that adults are facing; pedagogies that are customized to the adult brain; and flexible delivery models that fit in well with adult lifestyles. Adult learning is an important channel for readjusting skills to fit in the future of work, but it would benefit from a serious design rethink.

Adult learning programs come in many different forms. This section mainly focuses on three types that are particularly relevant to preparing adults for the changing labor markets: programs on adult literacy; skills training for wage employment; and entrepreneurship programs.

Worldwide, more than 2.1 billion working-age adults (ages 15–64) have low reading proficiency. In Sub-Saharan Africa, nearly 61 percent of workers are not proficient in reading; in Latin America and the Caribbean this share is 44 percent. In India, only 24 percent of 18- to 37-year-olds who drop out of school before completing the primary level can read.[28] A low-quality education also may lead to poor literacy skills (figure 4.6). In Bolivia, Ghana, and Kenya more than 40 percent of 19- to 20-year-olds with an upper secondary education score below the basic literacy level, compared with only 3 percent in Vietnam. This is a problem. Given the future of work, functional literacy is a survival skill. The economic and social cost of adult illiteracy to developing countries is estimated at more than US$5 billion a year.[29]

Even with basic literacy skills, many people leave school too early to thrive in work or life. Reasons may be economic or cultural constraints, the low quality of basic education, or both. In 2014 the dropout rate from a lower secondary general education was, on average, 27.5 percent in low-income countries and 13.3 percent and 4.8 percent in middle- and high-income countries, respectively.[30] It is difficult for early school leavers to find jobs or

FIGURE 4.6 In some economies, a large share of 19- to 20-year-olds have low literacy skills, despite completion of a secondary education

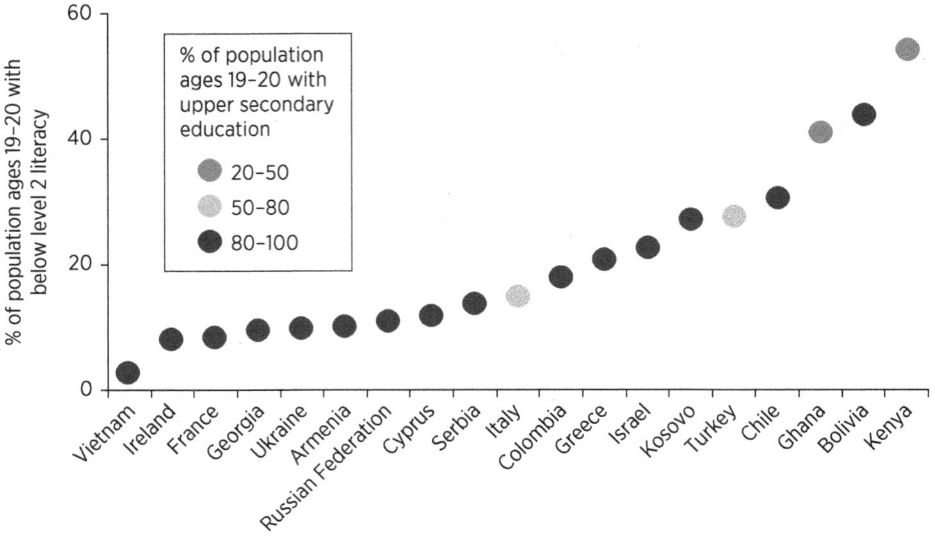

Source: WDR 2019 team.

Note: Data on Armenia, Bolivia, Colombia, Georgia, Ghana, Kenya, Kosovo, Serbia, Ukraine, and Vietnam: World Bank's STEP Skills Measurement Surveys; data on rest of economies: Programme for the International Assessment of Adult Competencies (PIAAC) data set. Tertiary education is merged with upper secondary education. STEP surveys are representative of urban areas. The PIAAC sample for the Russian Federation does not include the population of the Moscow municipal area.

pursue further education later in life without formal certification and training in skills. Similar constraints are also faced by many adults who stayed in school but received a poor-quality basic education.

Globally, some 260 million people ages 15–24 are out of school and out of work. A pool of unemployed adults is a political risk as well as an economic concern. At times, it leads to a wave of emigration, social unrest, or political upheaval. Insufficient economic opportunities for an increasingly educated population were a major catalyst of the 2010–11 Arab Spring. Changing demographics place additional pressures on the labor market. Many rich countries are trying to equip a smaller, older workforce with new skills for the changing nature of work, to sustain economic growth. Other countries with big youth cohorts are struggling with a low-skill labor force trapped in low-productivity jobs.

Adult learning programs update the skills and retool and improve the adaptability of older workers. India's Saakshar Bharat initiative, launched in 2009, seeks to provide 70 million adults with literacy. In Ghana, adult literacy programs have yielded labor market returns of more than 66 percent.[31] The Mexican National Institute for Adult Education has developed flexible modules to deliver education programs equivalent to a primary or secondary education. They are intended to give out-of-school individuals a second chance. Under the World Bank's Nepal: Adolescent Girls Employment Initiative, vocational training for women has increased employment outside of agriculture by 174 percent.[32] Argentina's Entra21 program is providing adult skills training and internships, resulting in 40 percent higher earnings for its participants.[33] Kenya's Ninaweza program is offering skills training to young women living in informal settlements in Nairobi. The program has led to a 14 percent increase in the likelihood of obtaining a job, higher earnings, and more self-confident participants.[34]

But many adult learning programs fail to generate a meaningful impact. Adult literacy programs often improve word recognition but fail to improve actual reading comprehension.[35] In Niger, an adult education program increased reading speed, but not to the level required for reading comprehension (the minimum reading speed for reading comprehension is one word every 1.5 seconds). Entrepreneurship programs often improve business knowledge, but they do not create employment. In Peru, training for female entrepreneurs improved business, but it did not generate a significant increase in employment. Vocational training for the unemployed often improves short-run earnings but not always long-run employment. The Dominican Republic's Juventud y Empleo (Youth and Employment) program improved noncognitive skills and job formality, but it did not increase employment. And Turkey's vocational training had no significant impacts on overall employment, and the positive effects on employment quality faded in the long term.

Even among successful adult learning programs, the costs are high. In Liberia, even though young women with access to job skills training enjoy higher monthly earnings—US$11 more than the comparison group—the cost of the program is US$1,650 per person.[36] Thus 12 years of stable effects must pass for the training program to recoup its costs. In Latin America, a long time is required for some programs to attain positive net present values if their benefits are sustained—for example, seven years for ProJoven (Program for Young People) in Peru and 12 years for Proyecto Joven (Young Project) in Argentina.[37] Adult learning is frequently just one expensive component of a comprehensive package, making it difficult to understand a program's cost-effectiveness. The Chilean Micro Entrepreneurship Support Program boosted self-employment by 15 percentage points in the short run, but it is not clear how much of this can be attributed to the 60-hour business training or the US$600 capital injection.[38]

The two main reasons for low effectiveness are a suboptimal design and an incorrect diagnosis. Adult brains learn differently—and that is not always factored into program design. Because the brain's ability to learn lessens with age, adult learning programs face a built-in challenge: acquiring knowledge when the brain is less efficient at learning. Advances in neuroscience suggest how to tackle this factor. An adult brain's ability to learn is significantly dependent on how much it is used. Adult learning programs have a better chance of success if lessons are integrated into everyday life.

In Niger, students who received instruction via their mobile phones as part of an adult education program achieved reading and math scores that were significantly higher than those who did not.

Adults face significant stress, which compromises their mental capacity—and that, too, is not always factored into program design. For adults, emotions are constantly mediated by the demands of family, child care, and work. These demands compete with the cognitive capacity required for learning. In India, sugarcane farmers were found to have a markedly diminished cognitive capacity when they were poorer (during preharvest) than when they were richer (during postharvest). Creating emotional cues linked to learning content—such as goal-setting—can be an effective strategy to increase adult learning. But behavioral tools are rarely integrated into adult learning programs.

Adults face specific socioeconomic constraints—and, again, these are not always factored into the design of adult learning programs. Adult learners have high opportunity costs in terms of lost income and lost time with their children, but programs often have inflexible and intensive schedules. In Malawi, participation in training resulted in a decline in personal savings for women at a rate nearly double that of men. Distance to training locations and lack of child care were significant barriers for women trying to complete vocational training programs in India. For adult literacy programs, dropout rates are often high, ranging from 17 percent in Niger to 58 percent in India.[39]

Low participation in adult learning programs is a sign that they are not always the answer. In Pakistan's Skills for Employability program, even among poor households who expressed interest in vocational skills, more than 95 percent did not enroll when given a voucher. Even when the government increased daily stipends and moved the training centers to villages, enrollment did not exceed 25 percent.[40] In Ghana, the demand for training by informal businesses is low because most managers do not see lack of skills as a constraint.

Three promising routes to more effective adult learning programs are better diagnosis and evaluation, better design, and better delivery.

For better diagnosis and evaluation, systematic data collection before program design will identify the most important constraints for the target population. This information is also useful for customizing skills training. Administrative data from India's massive National Rural Employment Guarantee Act program has offered powerful insights into local labor markets.

There is tremendous scope for improving the design of adult learning programs using insights from neuroscience and behavioral economics. Both practical exercises and visual aids are effective in adult learning because they help memory. Including motivational tools such as financial rewards, work experience, or frequent feedback have all been shown to boost adult learning. An experiment among young adults shows that offering rewards increases long-term performance gains after training.

As for delivery, flexible adult learning programs allow adults to learn at their convenience. In a voucher program for vocational training in Kenya, nearly 50 percent of women cited proximity to a training center as a determining factor in choosing a course.[41] Given competing demands on adults' time, training programs with short modules delivered through mobile applications are particularly promising. Delivering training programs via mobile phones better shields adult learners from potential stigma.

Adult learning programs are more successful when they are explicitly linked to employment opportunities. One popular way to do this is through apprenticeships or internships that link training to day-to-day experience and provide motivation through the promise of future economic returns. Skills training programs are more successful when the private sector is involved in developing the curriculum or training methods or in providing on-the-job training via internships or apprenticeships. Colombia's Jóvenes en Acción (Youth in Action) program combines classroom instruction with on-the-job training at private companies. The probability of formal employment and earnings rose in the short term and has been sustained in the long run. The program has also demonstrated strong education effects, with participants more likely to complete secondary school and to pursue higher education eight years after the training. The likelihood of their family members enrolling in tertiary education also has increased.

The success of adult learning programs may also depend on addressing multiple constraints at the same time. Combining training with cash or capital in some cases is a direct way to boost effectiveness. In Cameroon, 54,000 people who participated in a program that coupled training with financial assistance found employment.[42] Combining skills training with skills certificates, referral letters, and better information about job opportunities also may enhance effectiveness, especially for women. In Uganda, workers with more certifiable, transferrable skills have higher employment rates, higher earnings, and greater labor market mobility. A World Bank program in South Africa is attempting to improve job searches through peer support, text message reminders, and action planning.

Incorporating soft skills or sociobehavioral skills in training design has shown promise. In Togo, teaching informal business owners "personal initiative"—a mindset of self-starting behavior, innovation, and goal-setting—boosted the profits of firms by 30 percent two years after the program. This approach was much more effective than traditional business training. For factory workers in India, acquiring skills such as time management, effective communication, and financial management increased their productivity.

Notes

1. Krueger and Kumar (2004).
2. Cunningham and Villaseñor (2016); Deming (2017).
3. Hanushek et al. (2017).
4. World Bank's International Income Distribution Data Set.
5. Ederer et al. (2015).
6. World Bank (2015a, 2015b).
7. WDR 2019 team, based on World Bank's Enterprise Surveys, 2015–16.
8. Ebenehi, Rashid, and Bakar (2016).
9. World Bank (2018b).
10. Attanasio et al. (2014).
11. Oxford Policy Management (2009).
12. Aboud and Hossain (2011).
13. Martinez, Naudeau, and Pereira (2012).
14. Garcia, Heckman, and Ziff (2017).
15. Bidwell and Watine (2014).
16. Engle et al. (2011).
17. Nollenberger and Rodríguez-Planas (2015).
18. Hoddinott et al. (2008).
19. Gertler et al. (2014).
20. Black et al. (2017).
21. UNESCO (2015, 59).
22. Brinkman et al. (2017).
23. Psacharopoulos and Patrinos (2018).
24. Saavedra (2009).
25. Ferreyra et al. (2017).
26. Hasanefendic, Heitor, and Horta (2016).
27. Blom et al. (2016); StudyMalaysia (2016).
28. Kaffenberger and Pritchett (2017).
29. Cree, Kay, and Steward (2012).
30. Based on the "cumulative drop-out rate to the last grade of lower secondary general education" indicator published by the UNESCO Institute for Statistics. Data are available for 112 economies.
31. Blunch, Darvas, and Favara (2018).
32. Chakravarty et al. (2017).
33. J-PAL (2017).
34. Alvares de Azevedo, Davis, and Charles (2013).
35. Aker and Sawyer (2016).
36. Adoho et al. (2014).
37. Kluve (2016).
38. Martínez, Puentes, and Ruiz-Tagle (2018).
39. Aker and Sawyer (2016).
40. Cheema et al. (2015).
41. Hicks et al. (2011).
42. Haan and Serrière (2002).

References

Aboud, Frances E., and Kamal Hossain. 2011. "The Impact of Preprimary School on Primary School Achievement in Bangladesh." *Early Childhood Research Quarterly* 26: 237–46.

Adoho, Franck, Shubha Chakravarty, Dala T. Korkoyah, Jr., Mattias Lundberg, and Afia Tasneem. 2014. "The Impact of an Adolescent Girls Employment Program: The EPAG Project in Liberia." Policy Research Working Paper 6832, World Bank, Washington, DC.

Aker, Jenny, and Melita Sawyer. 2016. "Adult Learning in Sub-Saharan Africa: What Do and Don't We Know?" Background paper for Arias et al., forthcoming, "The Skills Balancing Act in Sub-Saharan Africa: Investing in Skills for Productivity, Inclusion, and Adaptability."

Alvares de Azevedo, Thomaz, Jeff Davis, and Munene Charles. 2013. "Testing What Works in Youth Employment: Evaluating Kenya's Ninaweza Program." Global Partnership for Youth Employment, Washington, DC.

Attanasio, Orazio P., Camila Fernández, Emla O. A. Fitzsimons, Sally M. Grantham-McGregor, Costas Meghir, and Marta Rubio-Codina. 2014. "Using the Infrastructure of a Conditional Cash Transfer Program to Deliver a Scalable Integrated Early Child Development Program in Colombia: Cluster Randomized Controlled Trial." *BMJ* 349 (September 29): g5785.

Bidwell, Kelly, and Loïc Watine. 2014. *Exploring Early Education Programs in Peri-urban Settings in Africa.* New Haven, CT: Innovations for Poverty Action.

Black, Maureen M., Susan P. Walker, Lia C. H. Fernald, Christopher T. Andersen, Ann M. DiGirolamo, Chunling Lu, Dana C. McCoy, et al. 2017. "Early Childhood Development Coming of Age: Science through the Life Course." *Lancet* 389 (10064): 77–90.

Blom, Andreas, Reehana Rana, Crispus Kiamb, Him Bayusuf, and Mariam Adil. 2016. "Expanding Tertiary Education for Well-Paid Jobs: Competitiveness and Shared Prosperity in Kenya." World Bank, Washington, DC.

Blunch, Niels-Hugo, Peter Darvas, and Marta Favara. 2018. "Unpacking the Returns to Education and Skills in Urban Ghana: The Remediation Role of Second Chance Learning Programs." Working paper, World Bank, Washington, DC.

Brinkman, Sally Anne, Amer Hasan, Haeil Jung, Angela Kinnell, and Menno Pradhan. 2017. "The Impact of Expanding Access to Early Childhood Education Services in Rural Indonesia." *Journal of Labor Economics* 35 (S1): 305–35.

Chakravarty, Shubha, Mattias Lundberg, Plamen Nikolov, and Juliane Zenker. 2017. "Vocational Training Programs and Youth Labor Market Outcomes: Evidence from Nepal." HCEO Working Paper 2017–056, Human Capital and Economic Opportunity Global Working Group, University of Chicago, July.

Cheema, Ali, Asim I. Khwaja, Muhammad Farooq Naseer, and Jacob N. Shapiro. 2015. "Skills Intervention Report: Results of First Round of Voucher Disbursement and Strategies for Improving Uptake." Technical Report, Punjab Economic Opportunities Program, Punjab Social Protection Authority, Lahore, Pakistan.

Cree, Anthony, Andrew Kay, and June Steward. 2012. "The Economic and Social Cost of Illiteracy: A Snapshot of Illiteracy in a Global Context." World Literacy Foundation, Melbourne, April.

Cunningham, Wendy, and Paula Villaseñor. 2016. "Employer Voices, Employer Demands, and Implications for Public Skills Development Policy Connecting the Labor and Education Sectors." *World Bank Research Observer* 31 (1): 102–34.

Deming, David J. 2017. "The Growing Importance of Social Skills in the Labor Market." *Quarterly Journal of Economics* 132 (4): 1593–1640.

Ebenehi, Amos Shaibu, Abdullah Mat Rashid, and Ab Rahim Bakar. 2016. "Predictors of Career Adaptability Skill among Higher Education Students in Nigeria." *International Journal for Research in Vocational Education and Training* 3 (3): 212–29.

Ederer, Peer, Ljubica Nedelkoska, Alexander Patt, and Silvia Castellazzi. 2015. "What Do Employers Pay for Employees' Complex Problem Solving Skills?" *International Journal of Lifelong Education* 34 (4): 430–47.

Engle, Patrice L., Lia C. H. Fernald, Harold Alderman, Jere R. Behrman, Chloe O'Gara, Aisha Yousafzai, Meena Cabral de Mello, et al. 2011. "Strategies for Reducing Inequalities and Improving Developmental Outcomes for Young Children in Low-Income and Middle-Income Countries." *Lancet* 378 (9799): 1339–53.

Ferreyra, María Marta, Ciro Avitabile, Javier Botero Álvarez, Francisco Haimovich Paz, and Sergio Urzúa. 2017. *At a Crossroads: Higher Education in Latin America and the Caribbean*. Directions in Development: Human Development Series. Washington, DC: World Bank.

Garcia, Jorge Luis, James J. Heckman, and Anna L. Ziff. 2017. "Gender Differences in the Benefits of an Influential Early Childhood Program." NBER Working Paper 23412, National Bureau of Economic Research, Cambridge, MA.

Gertler, Paul J., James J. Heckman, Rodrigo Pinto, Arianna Zanolini, Christel Vermeersch, Susan P. Walker, Susan M. Chang, et al. 2014. "Labor Market Returns to an Early Childhood Stimulation Intervention in Jamaica." *Science* 344 (6187): 998–1001.

Haan, Hans Christian, and Nicolas Serrière. 2002. *Training for Work in the Informal Sector: Fresh Evidence from West and Central Africa*. Occasional paper, International Training Centre, International Labour Organization, Turin, Italy, August.

Hanushek, Eric A., Guido Schwerdt, Simon Wiederhold, and Ludger Woessmann. 2017. "Coping with Change: International Differences in the Returns to Skills." *Economics Letters* 153 (April): 15–19.

Hasanefendic, Sandra, Manuel Heitor, and Hugo Horta. 2016. "Training Students for New Jobs: The Role of Technical and Vocational Higher Education and Implications for Science Policy in Portugal." *Technological Forecasting and Social Change* 113 (Part B): 328–40.

Hicks, Joan Hamory, Michael Kremer, Isaac Mbiti, and Edward Miguel. 2011. "Vocational Education Voucher Delivery and Labor Market Returns: A Randomized Evaluation among Kenyan Youth." Report for Spanish Impact Evaluation Fund, Phase II, World Bank, Washington, DC.

Hoddinott, John, John A. Maluccio, Jere R. Behrman, Rafael Flores, and Reynaldo Martorell. 2008. "Effect of a Nutrition Intervention during Early Childhood on Economic Productivity in Guatemalan Adults." *Lancet* 371 (9610): 411–16.

J-PAL (Abdul Latif Jameel Poverty Action Lab). 2017. "Skills Training Programmes." Cambridge, MA, August.

Kaffenberger, Michelle, and Lant Pritchett. 2017. "More School or More Learning? Evidence from Learning Profiles from the Financial Inclusion Insights Data." Background paper prepared for WDR 2018, World Bank, Washington, DC.

Kluve, Jochen. 2016. "A Review of the Effectiveness of Active Labour Market Programmes with a Focus on Latin America and the Caribbean." Research Department Working Paper 9, International Labour Office, Geneva, March.

Krueger, Dirk, and Krishna B. Kumar. 2004. "Skill-Specific Rather than General Education: A Reason for US-Europe Growth Differences?" *Journal of Economic Growth* 9 (2): 167–207.

Martínez, Claudia, Esteban Puentes, and Jaime Ruiz-Tagle. 2018. "The Effects of Micro-Entrepreneurship Programs on Labor Market Performance: Experimental Evidence from Chile." *American Economic Journal: Applied Economics* 10 (2): 101–24.

Martinez, Sebastian, Sophie Naudeau, and Vitor Pereira. 2012. "The Promise of Preschool in Africa: A Randomized Impact Evaluation of Early Childhood Development in Rural Mozambique." World Bank, Washington, DC.

Nollenberger, Natalia, and Núria Rodríguez-Planas. 2015. "Full-Time Universal Childcare in a Context of Low Maternal Employment: Quasi-Experimental Evidence from Spain." *Labour Economics* 36 (October): 124–36.

Oxford Policy Management. 2009. "Lady Health Worker Programme: External Evaluation of the National Programme for Family Planning and Primary Health Care: Summary of Results." Islamabad, Pakistan.

Psacharopoulos, George, and Harry Anthony Patrinos. 2018. "Returns to Investment in Education: A Decennial Review of the Global Literature." Policy Research Working Paper 8402, World Bank, Washington, DC.

Saavedra, Juan Esteban. 2009. "The Learning and Early Labor Market Effects of College Quality: A Regression Discontinuity Analysis." Working paper, University of Southern California, Los Angeles.

StudyMalaysia. 2016. "Technical and Vocational Education and Training (TVET) in Malaysia." October 12. https://www.studymalaysia.com/education/top-stories/technical-and-vocational-education-and-training-in-malaysia.

UNESCO (United Nations Educational, Scientific, and Cultural Organization). 2015. *Education for All Global Monitoring Report*. Paris: UNESCO.

World Bank. 2015a. "Armenia: Skills toward Employment and Productivity (STEP), Survey Findings (Urban Areas)." Washington, DC, January 31.

———. 2015b. "Georgia: Skills toward Employment and Productivity (STEP), Survey Findings (Urban Areas)." Washington, DC, January 31.

———. 2018a. *Unleashing the Potential of Education in the Middle East and North Africa*. Washington, DC: World Bank.

———. 2018b. *World Development Report 2018: Learning to Realize Education's Promise*. Washington, DC: World Bank.

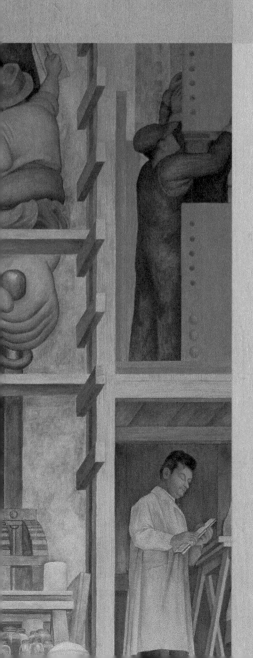

CHAPTER 5

Returns to work

earning does not end in school. Students who move into jobs have an opportunity to continue to accumulate human capital, but they face obstacles. For one thing, the emerging economies have a large informal sector. People working in this sector tend to be in low-productivity jobs that do not provide learning or stable sources of income. By creating the conditions for formal sector jobs, governments can offer better learning and income opportunities for the poor. Another obstacle is that women are often excluded from work. And yet another is that the poor in emerging economies are concentrated in rural areas in the agriculture sector. Raising their productivity is crucial to gaining human capital.

One of the fathers of labor economics quantified the payoffs from work and school. Before Jacob Mincer took on this subject in the 1970s, the common belief among his contemporaries was that luck determined a person's ability, which in turn determined payoffs. Mincer proved that earnings differentials are influenced by human capital investments that grow over the life cycle, initially in school and later at work. The payoff of such investments can be measured in terms of increased earnings, or "returns," stemming from an additional year spent in school or work. For example, Mincer found that for white males in nonfarm wage jobs an additional year of education increased earnings by 10.7 percent.[1]

Workers in emerging economies face lower payoffs to work experience than their counterparts in advanced economies (figure 5.1). In the Netherlands and Sweden, one additional year of work raises wages by 5.5 percent. In Afghanistan, the corresponding figure is 0.3 percent. A worker in an emerging economy is more likely than a worker in an advanced economy to engage in a manual occupation, where there is less scope for learning as well as a greater risk of automation. Moreover, compared with advanced economies, emerging economies have a poorly educated workforce. Advanced economies are often at the cutting edge of technology, and their workers tend to be highly educated, formally employed, and have access to a wide range of jobs filled with nonroutine, cognitive tasks. This may explain the higher returns to work in advanced economies than in emerging economies.

Comparing returns to work between manual and cognitive occupations reveals that an additional year of work in cognitive professions increases wages by 2.9 percent, whereas for manual occupations the figure is 1.9 percent.[2] Elementary occupations such as cleaners as well as skilled agricultural workers have the lowest returns. Professionals, managers, and technicians have the highest.

The workplace is a venue where skills can be acquired after school. Nevertheless, work is a complement to schooling, not a substitute for it. Global differences in school education explain much of the observed variation in earnings. One additional year spent in school produces, on average, the same increase in wages as spending four years at work. A worker would need to spend three years on the job in Germany, five years in Malawi, and eight years in Guatemala to match the effect of an extra year of schooling on wages. Policies that raise returns to work are likely to benefit more

FIGURE 5.1 High-income countries have higher returns to work experience than middle- and low-income countries

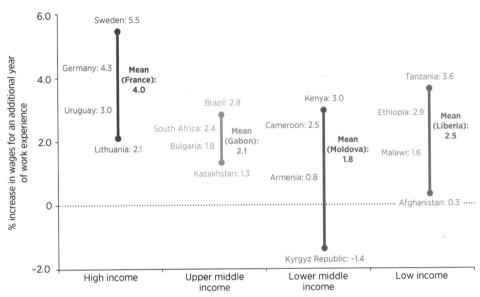

Source: WDR 2019 team, using household and labor force survey data from the World Bank's International Income Distribution Data Set.

Note: The figure provides estimates of the percentage increase in wages from an additional year of potential work experience across 135 economies by income level. For high-income economies, the mean is 4.0 percent. On average, then, an additional year of work experience increases monthly wages by 4.0 percent in high-income economies. The top and bottom economies for each income group are shown. Thus an additional year of experience raises monthly wages by 5.5 percent in Sweden, but only by 2.1 percent in Lithuania. The methodology follows previous work by categorizing years of experience into seven bins (Lagakos et al. 2018). Wage growth is estimated for each bin relative to the no-experience bin. The returns to experience are then calculated as an average of the seven bins, using a geometric mean. The top and bottom economy listed for each income group are ranked after the estimates account for income and life expectancy of the economy.

people in emerging economies because many of these workers are excluded from the school system.

Educated workers have greater scope for learning at work than uneducated workers. For each additional year of work experience, poorly educated workers have annual wage growth of 2 percent. By contrast, workers with high levels of education have annual returns to work of 2.4 percent. As a result, countries with poor schools face double jeopardy. First, young adults graduating from high school are not equipped with the skills to find work. Second, even if they find work they earn and learn less than more educated individuals.

Consider Jordan, a country with low returns to both education (5.9 percent) and experience (1.2 percent), together with below-average Programme for International Student Assessment (PISA) scores in math, science, and reading. A worker who completes secondary education in Jordan and one year at work would earn less than half of her counterpart in Germany. Moreover, by the time she accumulates 30 years of experience, the German worker's wage would already be at least five times higher than hers.

Informality

The informal economy is omnipresent in most emerging economies. Informal employment stands at more than 70 percent in Sub-Saharan Africa and 60 percent in South Asia and at more than 50 percent in Latin America. In Kenya, informal employment is a staggering 77.9 percent of total employment, one of the highest rates on the African continent. Almost 6 million businesses in Kenya's informal sector are unlicensed. And productivity is low in this sector: in emerging economies informal workers are, on average, only 15 percent as productive as formal workers.[3]

The informal sector is slow to change. Since 1999, India has seen its information technology sector boom; it has become a nuclear power; it has broken the world record in the number of satellites launched into outer space using a single rocket; and it has achieved an annual growth rate of 5.6 percent. Yet the size of its informal sector has remained at about 90 percent.[4] This pattern is not unique to India. Informal sectors in emerging economies are a fixture. In Madagascar, the percentage of nonagricultural informal workers increased from 74 percent in 2005 to 89 percent in 2012. In Nicaragua, informality rose from 72.4 percent in 2005 to 75 percent in 2010.[5]

Analyses based on the World Bank's International Income Distribution Data Set show that the returns to experience for a worker are higher in the formal sector than in the informal sector. For example, a year spent in the informal sector in Kenya raises income by only 2.7 percent a year. By contrast, workers in the formal sector in Kenya raise their income by 4.1 percent every year, which is approximately 1.5 times higher than for incomes in the informal sector. The difference is potent.

The disparity in the payoffs from work between formal and informal jobs is a global phenomenon. In Nepal, returns to experience are 1.4 times higher for formal wage workers over informal wage workers. In South Africa, they are 1.6 times higher in the formal sector than in the informal sector, and in India they are over twice as high as in the informal sector. In emerging economies, on average, the earnings increase for an additional year of work for informal wage workers is 1.4 percent, whereas it is 1.8 percent for formal wage workers.

The millions of informal businesses run by the poor are unlikely to make their owners rich. Typically, they have no paid staff and tend to be barely profitable. In Dakar, Senegal, 87 percent of firms with labor productivity below US$10,000 per worker are in the informal sector.[6] Informal firms are run by uneducated owners, serve low-income consumers, and use little capital—informal firms add only 15 percent of the value per employee of formal firms.[7] They also rarely move to the formal sector.

The poor manage to make a lot out of little, but the businesses they run are too small to raise their livelihoods. These enterprises do not provide a stable income stream, leaving the poor vulnerable to unexpected events. Yet they have no other options. The enterprises of the poor are a way to work when formal employment is unavailable.

Creating stable formal private jobs for the poor is an important policy goal. Stable jobs allow poor workers to make commitments to expenditures. Factory jobs dramatically improve the lives of the poor.[8] Improvements in infrastructure in towns and villages could encourage formal firms to establish themselves near poor workers. Although small-scale informal enterprises are unlikely to formalize and grow, the owners of informal firms may obtain formal jobs.

Countries that more heavily regulate their businesses have larger unofficial economies.[9] Mexico is a good illustration of what happens when a country streamlines its business regulation. In May 2002 Mexico began to implement its Rapid Business Opening System. The program simplified local business registration procedures and reduced the average number of days it takes to register a business from 30 to 2, the number of required procedures from 8 to 3, and the number of office visits required to register a business from 4 to 1. The Federal Commission for Improving Regulation organized the reform, coordinating with municipal governments because many business registration procedures are set locally in Mexico. Because of the reforms, informal business owners who were similar in profile to formal wage workers were 25 percent more likely to move into formal employment.[10] The evidence suggests that easing regulation encourages the transition from informal firm ownership to formal wage jobs.

At times, streamlining business has to be carried out in tandem with other policies. In Brazil, the Individual Micro-Entrepreneur Program, introduced in 2009, targeted entrepreneurs who employ at most one person. The program was designed to reduce the dimensions of formality costs—registration (entry) costs and the costs of remaining formal—by reducing monthly taxes and red tape. Reducing business registration costs while reducing taxes resulted in the formalization of existing informal firms. Industries eligible for tax reduction experienced a 5 percent increase in the number of formal firms. Halving monthly taxes led to a 2 percent increase in the registration rate for entrepreneurs, from a baseline rate of 20 percent.[11]

Governments can also use technology to reduce informality. The introduction of e-Payroll was an important factor in the reduction of nonagricultural informal employment in Peru from 75 percent in 2004 to 68 percent in 2012. Employers use e-Payroll to send monthly reports to the National Tax Authority on their workers, pensioners, service providers, personnel in training, outsourced workers, and claimants. E-payroll, which became available in 2008, resulted in the registration of about 300,000 new formal jobs in that year, after accounting for economic growth.[12]

Investments in human capital reduce informal employment. When young people are equipped with the right skills they are more likely to obtain a formal job. A youth training program in Santo Domingo, the capital of the Dominican Republic, targeted youths between the ages of 16 and 29 who did not attend school and were living in poor neighborhoods.[13] The program offered skills training courses that lasted 225 hours: 150 hours devoted to teaching a wide range of low skills for jobs such as administrative assistant, hair stylist, and mechanic, and 75 hours devoted to improving the soft skills of participants (mainly work habits and self-esteem). Courses were followed by a three-month internship in a private firm. Evaluation of the program revealed that skills investment in youth training has a significant impact on the probability of securing a formal job and on earnings in an urban labor market. And these gains last over time.

Working women

The mural *The Making of a Fresco Showing the Building of a City* by Mexican painter Diego Rivera (1886–1957) was chosen for the cover of this study. A communist, Rivera depicts a gigantic worker towering over bankers, architects, and artists. But only one woman appears among the 19 people in the mural. Although the status of women in the economy has improved since Rivera's time, a considerable gap remains in the economic opportunities available to women and men.

Some societies exclude women from work. Across the world, 49 percent of women over the age of 15 are employed, compared with 75 percent of men. Gender imbalances persist in positions of power. In less than a fifth of firms is a woman the top manager.[14] However, these numbers mask wide differences among countries. In Sweden, 61 percent of women are formally employed. In Italy, the figure is 40 percent. In India and Pakistan, only 25–27 percent of women are in the labor force. Generally, women work in less economically productive sectors and in occupations with potentially lower on-the-job learning opportunities.

The inclusion of women in formal economic activity depends on equal property rights. In ancient Greece, women could not inherit property rights, while in ancient Rome they had no political rights. In 1804 the Napoleonic Code stated that wives were under the purview of their fathers and husbands. Before 1870, married women in the United Kingdom had no right to claim property, and full ownership rights belonged to the husband. Although gender parity has improved around the world, major differences persist.

Women face sector-specific legal restrictions in obtaining jobs across many countries. Sixty-five economies restrict women from mining jobs; 47 impose restrictions in manufacturing; and 37 restrict women from construction jobs. Furthermore, in 29 of 189 economies women cannot work the same hours as men.

Men outnumber women in every occupation (figure 5.2). Only about a quarter of managers are women, and women are about 39 percent of professionals. Across occupations, women have a relatively higher presence in both clerical support and services and sales occupations (44 percent). The lowest presence is as plant and machine operators and assemblers, where women fill just 16 percent of such positions. Most female managers of formal firms in emerging economies are in the retail sector.

Women generally face lower payoffs from work experience (1.9 percent) than men (3.1 percent). In República Bolivariana de Venezuela, for each additional year of work, men's wages increase by 2.2 percent, compared

FIGURE 5.2 Men outnumber women across all broadly defined occupations

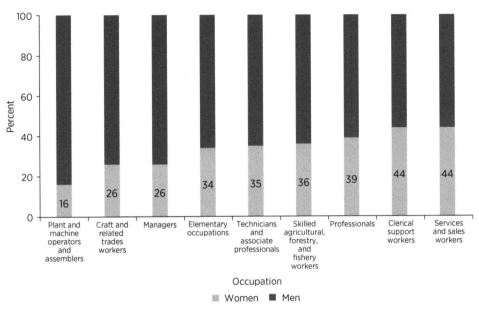

Source: WDR 2019 team, using household and labor force survey data from the World Bank's International Income Distribution Data Set.

with only 1.5 percent for women. The difference is even larger for countries such as Mali, where returns for men are 3.1 percent, but only 1.6 percent for women. A woman in Mali would have to accumulate almost two years more experience for every year her male coworker accumulates to earn the same wage increase. In Denmark, by contrast, this figure is 5 percent for both men and women.

There are many reasons for these different payoffs between men and women. Consider a working couple in Bangladesh who are contemplating conceiving their first baby. Bangladeshi laws do not prescribe paid or unpaid parental leave, so an equivalent job is not guaranteed for the mother after giving birth. Nursing mothers are not entitled to nursing breaks, and the law does not allow flexible or part-time schedules. Bangladesh's returns to work experience for women is 0.8 percent—almost half of the returns for men. By contrast, in Portugal, Spain, and Sweden—all countries with paid leave for both men and women—the returns to work experience are similar across genders.

Better information encourages change. As a response, the World Bank began the Women, Business and the Law project in 2008 to document the gender legal disparities for 189 economies. Removing legal restrictions for women is powerful. Simply mandating a nondiscrimination clause in hiring increases, in terms of gender, women's employment in formal firms by 8.6 percent.[15] Mandating paternity leave to encourage a more equitable distribution of child-rearing activities between men and women raises, on average, the proportion of women employed in formal firms by 6.8 percentage points.[16]

FIGURE 5.3 **More legal restrictions on women at work correspond with lower wages**

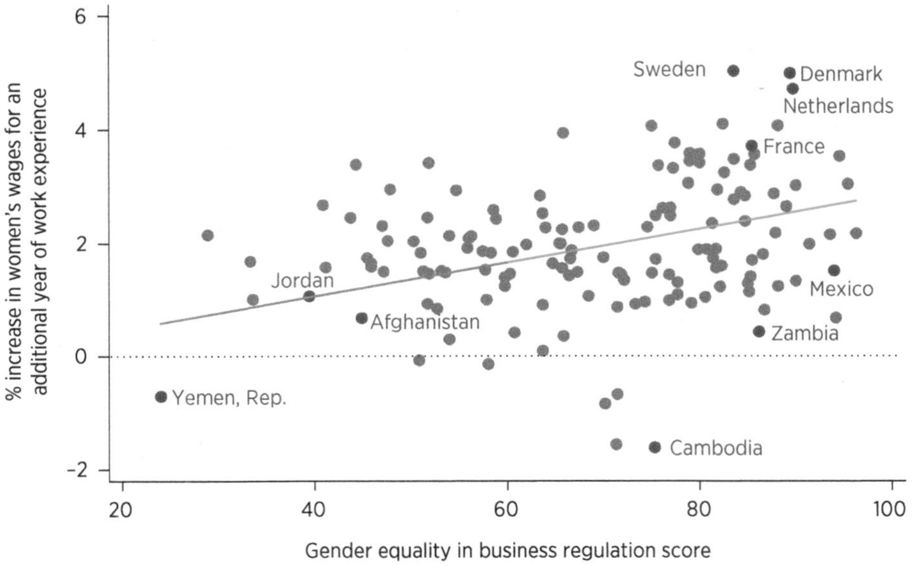

Sources: WDR 2019 team, based on World Bank (2018) and household and labor force survey data from the World Bank's International Income Distribution Data Set.

Note: The World Bank's Women, Business and the Law measure of gender legal equality scores economies based on whether they treat men and women differently. The higher the score, the greater is the gender legal equality.

The larger the number of legal restrictions women face, the lower is the payoff from work experience (figure 5.3). At one end of the spectrum, Denmark, France, the Netherlands, and Sweden have fewer legal gender restrictions and higher returns to work for women. In Afghanistan, Jordan, and the Republic of Yemen, where women and men are treated differently by law, the payoff from work experience for women is among the lowest. Increasing legal gender-specific restrictions discourages women from both owning and managing firms.[17] It is certainly possible that changes in the laws are not causing higher returns to work experience for women and that something else is. Nevertheless, laws are relatively easy to change and should be a natural first step.

Countries are reforming. Since changes to the family code in the Democratic Republic of Congo in 2016, a woman is allowed to register her business, open a bank account, apply for a loan, sign a contract, and register land without her husband's permission. Zambia's Gender Equity and Equality Act of 2015 prohibits gender discrimination in employment. Iraq guarantees workers returning after maternity leave a similar position and the same wage. China has increased its paid paternity leave. Afghanistan forbids sexual harassment in employment and education. Sixty-five countries made gender equality reforms between 2015 and 2017.

Reforming discriminatory laws and programs so they empower women by giving them access to training and assets improves their well-being. In Bangladesh, poor women generally work as maids or agricultural workers, and wealthy women rear livestock. A nationwide program changed lives by providing poor women with livestock as well as skills training and advice on their legal, social, and political rights. Earnings of many of the women in the program rose, the value of their livestock increased, they accumulated business assets, and they were more likely to own land. These improvements lasted seven years after the program.[18] A similar program in Uganda provided adolescent girls with vocational training, together with information on sexual health and reproduction to reduce early pregnancy. Four years after the program, women were more likely to be engaged in income-generating activities.[19]

Liberia launched the Economic Empowerment of Adolescent Girls and Young Women project in 2009. It seeks to provide young girls with both in-classroom training—focusing on life and technical skills highly demanded in the market—and follow-up job placement support (to either enter a paying job or start a new business). The program significantly improved participants' lives: employment and earnings increased by 47 and 80 percent, respectively; participating women saved US$35 more than a control group over a 14-month period; and their self-confidence, life satisfaction, and social abilities improved. Households with participating women improved their food security by increasing the consumption of high-value proteins, while decreasing the likelihood of food shortages.[20]

Working in agriculture

Agriculture remains the main economic sector in low-income countries, especially in rural areas, even as the number of jobs it supports falls as economies develop. In 2017 agriculture accounted for 68 percent of employment in low-income economies. Improving agricultural incomes is therefore an effective way of reducing poverty.[21] However, the combined forces of automation and open trade work against agricultural employment in developing countries. Meanwhile, capital-intensive agriculture in advanced economies may be reducing the demand for imports from developing economies.

The result is faster urbanization in Africa and South Asia, where the challenges of moving to cities proliferate. On the one hand, earnings could rise: in emerging economies an additional year of city work experience is worth a 2.2 percent increase in pay. The returns to work in urban areas are 1.7 times more than in rural areas where agriculture is predominant, or a premium of 70 percent. This reflects a global pattern. In Indonesia and Mexico, the returns to work are 50 percent higher in urban areas than in rural areas, and in China, India, and Vietnam the payoffs are double.

On the other hand, opportunities in the city can be limited. There, workers typically require some level of education to access most of the better jobs. In several developing economies, stringent workplace regulation deters firms from employing less productive workers, pushing them into the informal economy.[22]

The constraints faced by the poor in moving to cities have been well documented. In India, workers in Odisha give at least two reasons for not staying in the city.[23] First, there is no housing—the poorest often squeeze themselves into swamps or slums right next to refuse dumps. By contrast, villages offer more open, greener, and quieter spaces. Second, those moving their families to the city face considerable risks. If their children fall sick, the health care is better in the city, but will anyone lend them money if they are in need? The connections developed in villages serve as safety nets for the vulnerable lives of the poor.

To reduce poverty, governments may be tempted to move poor workers from villages that mainly rely on agriculture to cities to raise the overall payoffs for work experience in the economy. However, this movement is unlikely to do much to narrow the payoff gap between emerging and advanced economies. Studies conducted in Indonesia and Kenya have found that improvements in rural areas are necessary to narrow that gap.[24]

Between the bustling cities and the subsistence agriculture–oriented villages lie secondary towns. They play a special role in facilitating the transition of rural workers to off-farm employment, much of it related to agriculture. Secondary towns occupy an important space between villages and cities, enabling movement up and down the value chain. The experiences of Tanzanian migrants confirm this, highlighting the role that secondary towns play in facilitating the transition out of agriculture.[25] In the early stages of development, growing secondary towns may do more to alleviate rural poverty than cities. But in later stages of development, the cities take over.

As economies develop, agricultural productivity rises, unlike productivity in the informal sector. But the challenges facing farmers in emerging economies are many, and governments play an important part in raising productivity. Smallholders have limited access to agricultural inputs such as fertilizers and machinery, as well as services that increase their productivity; they are not integrated into value chains. Value chain development allows farmers to capture the urban demand for higher-value agricultural products such as dairy, meat, fruits, and vegetables. Poverty reduction is faster when agriculture shifts from raising staple crops to nonstaple crops. Such a step requires raising staple crop productivity well beyond the current levels in Sub-Saharan Africa. Policy makers are making progress in some areas. Examples are programs that transfer knowledge, and initiatives that exploit digital technologies to increase access to input, output, and capital markets.

Training farmers in the best farming techniques has been shown to raise productivity. Some projects expand training programs or collaboration to improve the exchange of information. At times, this has been combined with increasing access to finance or the inputs required for agriculture as an impetus for improving agricultural productivity. Providing cooperatives with resources improves linkages between agribusinesses along the value chain. JD Finance, the fintech arm of JD.com, a leading Chinese e-commerce platform, has been providing farmers with microcredits. For example, the Integrated Growth Poles Project in Madagascar, which provided farmers with training on improved cocoa processing practices and business management skills, resulted in beneficiaries seeing an average increase in net revenue of 47 percent. In Afghanistan, farmer field schools, which are part of the National Horticulture and Livestock Project, tripled the income of some participants. They have also been successful in East Africa.[26] Government initiatives to link farmers with producer organizations, agribusiness purchasers, and financial institutions in the sorghum sector in northern Cameroon had similar effects.

Agricultural training can be improved. One way is to activate social ties in villages to encourage peer learning. A recent study of rural female farmers in Uganda concluded that encouraging them to compete resulted in greater learning in training sessions.[27] Agricultural extension services can be improved through low-cost videos that leverage the knowledge and participation of local communities.

Mechanization has in the past failed to take a foothold in Sub-Saharan Africa, opening the door to skepticism about ambitious predictions of the technological transformation of agriculture. Yet, thanks to new information and communications technologies, there are signs that mechanization is happening. Instantaneous measurements allow farmers to make better decisions. Drones, aerial images from satellites, and soil sensors improve measurements and crop monitoring. Detailed information enables farmers to decide how much fertilizer and irrigation are appropriate.

Mobile technology in Kenya is reducing the administrative and assessment costs of crop insurance schemes. A good illustration is the app Kilimo Salama (Swahili for safe farming), which became ACRE Africa in 2014. The seller activates the insurance policy by scanning a product-specific bar code with a camera phone, entering the farmer's mobile number, and connecting the farmer to the local weather station. Thirty solar-powered weather stations automatically monitor the weather. The farmer receives a text message confirming the insurance policy. Indemnity payments are made through the M-Pesa platform. By 2017 over a million farmers in Kenya, Rwanda, and Tanzania were insured under this project.

Orchard farmers in Kastamonu Province in Turkey must contend with pests and frost. The government, in collaboration with international donors, established five mini-meteorological stations in rural areas throughout the province, along with 14 reference farms to measure rain, temperature, and pest cycles. Producers are regularly updated by text message, allowing them to react to the prevailing local conditions. Costs fell dramatically for producers in the first two years of the scheme. Pesticide applications dropped by 50 percent.

For farmers to profit from increased agricultural productivity, they need access to markets, both at home and abroad. Export-oriented agriculture in northern and central Mexico provides on-farm job opportunities for millions of farmers and for many others off-farm engaged in agrifood processing and packaging activities. Alquería, the third-largest dairy company in Colombia, is expanding exports—the 13,000 small dairy farmers from which Alquería sources raw milk will all benefit from the increased demand overseas. In addition to streamlining export processes, improving the trade logistics infrastructure, and increasing food safety compliance capacity, governments can facilitate exports through exporter training and marketing assistance.

For example, the Vietnamese government works with industry organizations to deliver coordinated branding campaigns for tea, coffee, and cashew nuts.

When farmers' crops finally go to market, many of those in emerging economies do not know whether they are getting the best prices. In Uganda, TruTrade is an example of a digital technology that is bridging this information gap. TruTrade connects smallholders to buyers while raising quality and transparency and creating an atmosphere of trust. It uses online applications to allow price-setting and to track the movements of produce and payments. Farmers receive good prices and reliable access to markets. Traders build relationships as trusted providers, thereby growing their businesses.

Work is the next venue for human capital accumulation after school. Poorer economies have much to do because they lag behind advanced economies in the returns to work. Governments can raise those returns by increasing formal jobs for the poor, enabling women's economic participation, and expanding agricultural productivity in rural areas. Formal jobs create more learning opportunities. Empowering women will raise the stock of human capital in the economy. And expanding agricultural productivity in rural areas will provide better work opportunities for the poor. Jobs that generate and build skills will prepare workers for the future.

Notes

1. Mincer (1974).
2. Levin et al. (2018).
3. La Porta and Shleifer (2014).
4. Kanbur (2017).
5. International Labour Organization's ILOSTAT database.
6. Benjamin and Mbaye (2012).
7. La Porta and Shleifer (2014).
8. Foster and Rosenzweig (2008).
9. Djankov et al. (2002).
10. Bruhn (2013).
11. Rocha, Ulyssea, and Rachter (2018).
12. FORLAC (2014).
13. Ibarrarán et al. (2018).
14. Islam et al. (2018).
15. Amin and Islam (2015).
16. Amin, Islam, and Sakhonchik (2016).
17. Islam, Muzi, and Amin (2018).
18. Bandiera, Burgess, et al. (2017).
19. Bandiera, Buehren, et al. (2017).
20. Adoho et al. (2014).
21. Christiaensen, Demery, and Kuhl (2011).
22. Divanbeigi and Saliola (2017).
23. Banerjee and Duflo (2011).
24. Hicks et al. (2017).
25. Ingelaere et al. (2018).
26. Davis et al. (2012); Larsen and Lilleør (2014).
27. Vasilaky and Islam (2018).

References

Adoho, Franck, Shubha Chakravarty, Dala T. Korkoyah, Jr., Mattias Lundberg, and Afia Tasneem. 2014. "The Impact of an Adolescent Girls Employment Program: The EPAG Project in Liberia." Policy Research Working Paper 6832, World Bank, Washington, DC.

Amin, Mohammad, and Asif Islam. 2015. "Does Mandating Nondiscrimination in Hiring Practices Influence Women's Employment? Evidence Using Firm-Level Data." *Feminist Economics* 21 (4): 28–60.

Amin, Mohammad, Asif Islam, and Alena Sakhonchik. 2016. "Does Paternity Leave Matter for Female Employment in Developing Economies? Evidence from Firm-Level Data." *Applied Economics Letters* 23 (16): 1145–48.

Bandiera, Oriana, Niklas Buehren, Robin Burgess, Markus P. Goldstein, Selim Gulesci, Imran Rasul, and Munshi Sulaiman. 2017. "Women's Empowerment in Action: Evidence from a Randomized Control Trial in Africa." Working paper, World Bank, Washington, DC.

Bandiera, Oriana, Robin Burgess, Narayan Das, Selim Gulesci, Imran Rasul, and Munshi Sulaiman. 2017. "Labor Markets and Poverty in Village Economies." *Quarterly Journal of Economics* 132 (2): 811–70.

Banerjee, Abhijit Vinayak, and Esther Duflo. 2011. *Poor Economics: A Radical Rethinking of the Way to Fight Global Poverty.* Philadelphia: PublicAffairs.

Benjamin, Nancy, and Ahmadou Aly Mbaye. 2012. *The Informal Sector in Francophone Africa: Firm Size, Productivity, and Institutions.* With Ibrahima Thione Diop, Stephen S. Golub, Dominique Haughton, and Birahim Bouna Niang. Africa Development Forum Series. Washington, DC: Agence Française de Développement and World Bank.

Bruhn, Miriam. 2013. "A Tale of Two Species: Revisiting the Effect of Registration Reform on Informal Business Owners in Mexico." *Journal of Development Economics* 103 (July): 275–83.

Christiaensen, Luc, Lionel Demery, and Jesper Kuhl. 2011. "The (Evolving) Role of Agriculture in Poverty Reduction: An Empirical Perspective." *Journal of Development Economics* 96 (2): 239–54.

Davis, Kristin E., Ephraim Nkonya, Edward Kato, Daniel Ayalew Mekonnen, Martins Odendo, Richard Miiro, and Jackson Nkuba. 2012. "Impact of Farmer Field Schools on Agricultural Productivity and Poverty in East Africa." *World Development* 40 (2): 402–13.

Divanbeigi, Raian, and Federica Saliola. 2017. "Regulatory Constraints to Agricultural Productivity." Policy Research Working Paper 8199, World Bank, Washington, DC.

Djankov, Simeon, Rafael La Porta, Florencio Lopez-de-Silanes, and Andrei Shleifer. 2002. "The Regulation of Entry." *Quarterly Journal of Economics* 117 (1): 1–37.

FORLAC (Programme for the Promotion of Formalization in Latin America and the Caribbean). 2014. "Trends in Informal Employment in Peru: 2004–2012." Notes on Formalization, FORLAC, Regional Office for Latin America and the Caribbean, International Labour Organization, Lima.

Foster, Andrew D., and Mark R. Rosenzweig. 2008. "Economic Development and the Decline of Agricultural Employment." In *Handbook of Development Economics,* vol. 4, edited by T. Paul Schultz and John A. Strauss, 3051–83. Handbooks in Economics Series 9. Amsterdam: North-Holland.

Hicks, Joan Hamory, Marieke Kleemans, Nicholas Y. Li, and Edward Miguel. 2017. "Reevaluating Agricultural Productivity Gaps with Longitudinal Microdata."

NBER Working Paper 23253, National Bureau of Economic Research, Cambridge, MA.

Ibarrarán, Pablo, Jochen Kluve, Laura Ripani, and David Rosas Shady. 2018. "Experimental Evidence on the Long-Term Impacts of a Youth Training Program." *ILR Review* (April 89). http://journals.sagepub.com/doi/10.1177/0019793918768260.

Ingelaere, Bert, Luc Christiaensen, Joachm De Weerdt, and Ravi Kanbur. 2018. "Why Secondary Towns Can Be Important for Poverty Reduction: A Migrant Perspective." *World Development* 105 (May): 273–82.

Islam, Asif, Isis Gaddis, Amparo Palacios-Lopez, and Mohammad Amin. 2018. "The Labor Productivity Gap between Female- and Male-Managed Firms in the Formal Private Sector." Policy Research Working Paper 8445, World Bank, Washington, DC.

Islam, Asif, Silvia Muzi, and Mohammad Amin. 2018. "Unequal Laws and the Disempowerment of Women in the Labor Market: Evidence from Firm-Level Data." *Journal of Development Studies* (July 2). https://www.tandfonline.com/doi/abs/10.1080/00220388.2018.1487055?journalCode=fjds20.

Kanbur, Ravi. 2017. "Informality: Causes, Consequences and Policy Responses." *Review of Development Economics* 21: 939–61.

Lagakos, David, Benjamin Moll, Tommaso Porzio, Nancy Qian, and Todd Schoellman. 2018. "Life-Cycle Wage Growth across Countries." *Journal of Political Economy* 126 (2): 797–849.

La Porta, Rafael, and Andrei Shleifer. 2014. "Informality and Development." *Journal of Economic Perspectives* 28 (3): 109–26.

Larsen, Anna Folke, and Helene Bie Lilleør. 2014. "Beyond the Field: The Impact of Farmer Field Schools on Food Security and Poverty Alleviation." *World Development* 64 (December): 843–59.

Levin, Victoria, Carla Solis Uehara, Marcela Gutierrez Bernal, Alexandria Valerio, and Magdalena Bendini. 2018. "The Adult Literacy Crisis: Lessons from Adult Skills Surveys." Background paper prepared for WDR 2019, World Bank, Washington, DC.

Mincer, Jacob. 1974. *Schooling, Experience, and Earnings*. Vol. 2 of *Human Behavior and Social Institutions*. Cambridge, MA: National Bureau of Economic Research; New York: Columbia University Press.

Rocha, Rudi, Gabriel Ulyssea, and Laísa Rachter. 2018. "Do Lower Taxes Reduce Informality? Evidence from Brazil." *Journal of Development Economics* 134 (September): 28–49.

Vasilaky, Kathryn N., and Asif Islam. 2018. "Competition or Cooperation? Using Team and Tournament Incentives for Learning among Female Farmers in Rural Uganda." *World Development* 103: 216–25.

World Bank. 2018. *Women, Business and the Law 2018*. Washington, DC: World Bank.

CHAPTER 6

Strengthening social protection

Otto von Bismarck, Germany's chancellor in the late 19th century, is widely credited with inventing social insurance—benefits for workers in the formal sector financed by dedicated taxes on wages. What is less known, however, is that this model was Bismarck's plan B. The chancellor's original intention was to create a system of pensions financed by taxes on tobacco. When this plan failed, Bismarck resorted to wage-based, contributory financing.

The Bismarckian model has served many countries well. However, in a range of developing countries the model has remained mostly aspirational because of the large size of their informal economies. As a result, many workers lack social protection. In low-income countries, of those in the poorest quintile, only 18 percent are covered by social assistance and 2 percent by social insurance. The corresponding rates increase to 77 and 28 percent in upper-middle-income settings.

This chapter outlines how three main components of social protection systems—a guaranteed social minimum (with social assistance at its core), social insurance, and labor market regulation—can manage labor market challenges (figure 6.1).[1] A social minimum includes the set of social assistance programs that provide financial support to a large share of the population, or even all of it. Already spurred by equity concerns, expanded social assistance is underscored by the growing risks in labor markets and the importance of providing adequate support, no matter how a person engages in those markets. A guiding principle for strengthening social assistance is progressive universalism. The aim of this approach is to expand coverage while giving priority to the poorest people. This bottom-up expansion occurs while navigating the fiscal, practical, and political trade-offs that incremental levels of coverage involve.

FIGURE 6.1 Social protection and labor regulation can manage labor market challenges

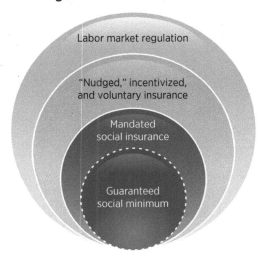

Labor market regulation

"Nudged," incentivized, and voluntary insurance

Mandated social insurance

Guaranteed social minimum

Source: WDR 2019 team.

Social assistance should be complemented with insurance that does not fully depend on formal wage employment. An arrangement of this nature would provide basic universal coverage, subsidizing premiums for the poor and topping up social assistance. Mandatory earnings-based contributions would also be necessary. This mandate would apply, at least initially, only to formal workers. A lighter mandate could attract greater compliance. Additional insurance could be achieved through voluntary saving schemes "nudged" by the state. Disentangling redistribution from savings would reduce labor costs. This change may also reduce incentives to replace workers with robots.

Taken together, the expanded coverage of social assistance and the provision of subsidized social insurance imply a stronger role for governments. For example, the desirable level of spending for a societal minimum in developing countries may, in many cases, be significantly higher than average expenditures on social assistance—currently at 1.5 percent of gross domestic product (GDP). Progressive universalism calls for gradual expansion in line with the prevailing fiscal space.

Enhanced social assistance and insurance reduce the burden on labor regulation of having to deal with risk management. As people become better protected through enhanced social assistance and insurance systems, labor regulation could, where appropriate, be made more flexible to facilitate movement between jobs. For example, income support for the unemployed could be provided by unemployment benefits rather than by severance pay.

Lower labor costs improve the adaptability of firms to the changing nature of work, while encouraging greater formal employment, especially of new entrants into the labor market and low-skill workers. Informal workers may also be better protected. However, a proper balance should be maintained between regulation and job creation. Complementary support for learning new skills, as well as new arrangements for strengthening the voice of workers, become even more important. Effective representation of both formal and informal sector workers ensures that the "security" element of "flexicurity" is preserved.

Social assistance

"All poor people should have the alternative . . . of being starved by a gradual process in the house, or by a quick one out of it." With these words, Charles Dickens's *Oliver Twist* provides a vivid depiction of social assistance practices in the 19th-century United Kingdom. The government, as codified in the Poor Laws of 1601 and 1834, established harsh criteria for accessing social assistance. The laws also influenced thinking about social assistance for centuries. It was only 75 years ago that the "Beveridge Report," with its recommendations embedded in the 1948 National Assistance Act, marked the end of the era evoked by Dickens.

In subsequent decades, social assistance began to spread to developing countries. Trends in social assistance attest to significant global progress. Analysis based on 142 countries, including those in the World Bank's database on social protection—the Atlas of Social Protection Indicators of Resilience and Equity, ASPIRE—shows that 70 percent have unconditional cash transfers in place, and 43 percent rely on conditional cash transfers. Meanwhile, 101 countries have old-age social pensions.[2]

Developing countries are continually expanding their social assistance programs. For example, the coverage of the national conditional cash transfer scheme in Tanzania increased from 0.4 percent of the population in 2013 to 10 percent in 2016. An equal level of coverage has been achieved by the Productive Safety Net Program in Ethiopia. In the Philippines, about 20 percent of the population is served by the Pantawid program, and in South Africa, by the Child Support Grant. Overall, the cumulative coverage of social assistance is 40.1 percent of the 5.1 billion people represented in surveys included in ASPIRE.

Social assistance works on many levels. Empirical studies have shown that cash transfers are spent on food, health care, education, and other desirable goods. Transfers are associated with improvements in the human capital of current and future generations. A systematic review of 56 cash transfer programs found significant advances in school enrollment rates, test scores, cognitive development, food security, and use of health facilities.[3] In Mexico, the Prospera conditional cash transfer program improved motor skills, cognitive development, and the receptive language of children from age 24 to 68 months. In Kenya, secondary school enrollment increased by 7 percent for children in the Orphans and Vulnerable Children program. These gains are usually largest for the poorest, rural dwellers, girls, and ethnic minorities. Cash transfers reduce stress and depression, increase mental bandwidth, and foster more involved parenting.[4]

Social assistance programs also affect household assets and livelihoods. Evaluations in Africa found that livestock ownership increased, on average, by 34 percent and ownership of durable goods by 10 percent.[5] Programs increasingly reinforce livelihood effects by adding elements of awareness-raising on nutritional risks, financial inclusion, entrepreneurship training, and asset transfers. In other words, social assistance, especially income support plus interventions, has often raised productivity and resilience among informal workers.

In advanced economies, social assistance faces the challenge of low uptake among eligible beneficiaries. In the European Union, only about 60 percent of social benefits are claimed.[6] This challenge stems from a lack of awareness of benefits, misunderstanding of eligibility rules, the stigma associated with assistance, bureaucratic obstacles, and the opportunity costs of accessing benefits.

In some middle-income countries with high levels of coverage, policy makers have weighed the possibility of targeting by excluding the rich instead of selecting beneficiaries from the bottom. This approach is often considered in the context of large-scale energy and food subsidy reforms. The political viability of such a proposition may then depend on how the middle class and various interest groups are set to benefit from (and in part pay for) the program as part of a wider social contract.

Where deprivation is widespread, households across the income distribution face similar levels of need. Continuity in welfare distribution may contrast with sharp, somewhat arbitrary, measures of poverty or eligibility criteria. For example, in some middle-income countries people living on US$6 a day, or just above the poverty line, face a 40 percent probability of falling back into poverty.[7] Poverty is often dynamic: in Africa, a third of the population is persistently poor, while another third moves in and out of poverty.[8]

These facts suggest the need for broader and more permanent coverage than most social assistance programs currently provide. Although more universal approaches are desirable, the specific shape of this social minimum faces technical, budgetary, and political challenges. Universal approaches typically reduce or eliminate hurdles around program fragmentation, eligibility determination, and social tensions, but they require significant additional resources. Expanding social assistance should proceed at the same pace as the mobilization of required resources. The choice of larger or smaller tax transfer policies has distributional effects as well as diverse bases of political support.

As part of the expansion options, a universal basic income (UBI) is being hotly debated. This tool enshrines the notion of building a guaranteed social minimum through a single program with three design features. First, the program is aimed at every individual, independent of income or employment status. Second, participants do not have to fulfill any conditions or reciprocal co-responsibilities. Third, assistance is provided in the form of cash instead of in-kind transfers and services (figure 6.2).

FIGURE 6.2 A universal basic income (UBI) has limited targeting, no conditions, and is paid in cash to recipients

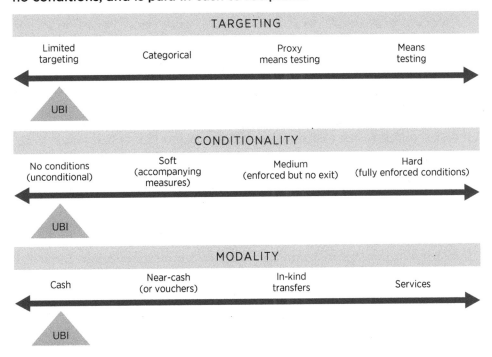

Source: WDR 2019 team.

Note: Each of the three sections of the figure includes illustrative alternative design traits.

A UBI is not an alternative to health, education, or other social services. The program may supplement current social assistance programs and more likely replace some of the programs pursuing income support functions. It may be designed with different objectives in mind, from poverty reduction to ensuring a livable income. The discussion here focuses on poverty reduction. Although a UBI provides the entire population with the same level of benefits, the money may be recovered from the rich through, for example, a progressive income tax.

In early 2018 India's then chief economic adviser, Arvind Subramanian, predicted enthusiastically, "I can bet . . . within the next two years, at least one or two [Indian] states will implement UBI."[9] But little is known about how a UBI works in practice. Only one country, Mongolia, has had such an initiative covering the entire population. However, the program lasted only two years (2010–12) before being dismantled because of fiscal constraints (when mineral prices collapsed, so did the scheme). The Islamic Republic of Iran had a similar program for one year: in 2011 energy subsidies were replaced by cash transfers to 96 percent of the population.

Local variants of a UBI are in place in a range of resource dividend schemes. In the United States, the Alaska Permanent Fund is designed to redistribute oil revenues to all state residents. In 2016 the fund distributed about US$2,000 each to 660,000 Alaskans. Several small-scale schemes and experiments are under way or being explored in China, Kenya, the Netherlands, Scotland, and the United States. Although they are called UBI programs, they are in some instances versions of targeted programs.

The fiscal implications of a UBI could be significant. New analysis has estimated the costs of providing such a program in four European countries. UBI transfers were set equal to those of existing cash transfer programs.[10] According to the results, the additional cost of a UBI varies significantly— 13.8 percent of GDP in Finland, 10.1 percent in France, 8.9 percent in the United Kingdom, and 3.3 percent in Italy. To cover the additional costs, two funding sources were identified: taxing UBI transfers alongside other incomes, and abolishing existing tax allowances. In Finland and Italy, these measures were more than adequate to cover the additional costs of a UBI. In France, those revenues almost offset the cost of such a program. In the United Kingdom, taxing cash benefits and eliminating tax allowances were not enough to cover the UBI.

Simulations from developing countries also point to significant additional spending for a UBI. In a handful of emerging economies, a UBI set at 25 percent of median income would cost about 3.8 percent of GDP.[11] By comparison, low- and middle-income countries spend, on average, 1.5 percent of GDP on social assistance. In India, the government estimates that a quasi-UBI excluding the top 25 percent could be largely paid for by replacing existing schemes.[12] Although those schemes account for about 5 percent of GDP, the results of the simulation have been contested.[13] Elsewhere, other simulations are providing further evidence. The cost of a UBI for adults set at the average poverty gap level ranges from 9.6 percent of GDP in low-income countries to 3.5 percent of GDP in upper-middle-income countries. If transfer amounts are lower—for example, set at the average level of current benefits—the costs would fall considerably (but would have less impact). Whether a UBI provides sufficient funds to close the poverty gap or equates to current transfer levels, the cost of the scheme would nearly double if intended for an entire population instead of for adults only (figure 6.3).

A UBI would generate winners and losers among the population. Its effects would depend on how the program is financed; whether existing targeted programs would be replaced and which ones; the performance of existing schemes; current tax structures; the size of UBI transfers; and the profile of people receiving it.

FIGURE 6.3 **The cost of a universal basic income climbs as the income level of countries decreases**

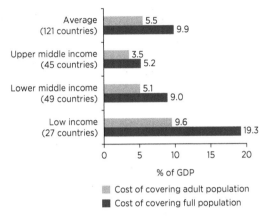

Sources: WDR 2019 team, based on World Bank's World Development Indicators (database) and PovcalNet and United Nations' World Population Prospects.

Note: GDP = gross domestic product.

Estimates for a selection of developing countries that simulate replacing some existing schemes with a UBI found significant distributional effects. In Nepal, most people would gain from such a program. In Indonesia, a UBI providing the same average amount of benefits as current programs would make most of the population better off, but about 40 percent of the poor would receive less. Under the same scenario, simulations suggest that a UBI in South Africa would make most of the elderly and the poor worse off. A similar negative effect on about 40 percent of senior citizens would be observed in Chile.

A recurring concern around the UBI is the risk of work disincentives. In theory, a UBI has only an income effect: the fact that the program's benefits are delinked from earnings or other income may suggest there is no substitution effect. However, the available evidence confirms that both a UBI and other forms of social assistance have a limited impact on work incentives. A study of the Alaskan dividend program shows no impact on employment. Instead, it finds that part-time employment increases by 1.8 percentage points (equivalent to a 17 percent increase in employment).[14] A study of the Iranian quasi-UBI program found that it did not affect the overall labor supply.[15]

An important debate is whether a program that guarantees jobs would serve as a better alternative to a UBI. India's National Rural Employment Guarantee Act offers 100 days of work every year at the minimum wage. UBI proponents contest such a public works approach by asserting that the right to income should precede the right to work. Conversely, it is contended that the right to work rests on the premise that anyone who wants work could be offered a job, thereby conferring on work a societal value. Those favoring job schemes also point to the range of productive and socially valuable activities that could be implemented beyond labor-intensive tasks such as social care services. A UBI may be an alternative to public works when their overwhelming function is mere income support. However, when more meaningful activity is envisioned, public works emerge as a complementary instrument for those who are fit and able to work. The concept of "participation income" is a hybrid between a UBI and public works. It envisions providing universal cash transfers tied to some form of civil engagement.

A UBI could generate more efficiency by reducing program fragmentation. Most countries operate a complex mosaic of social assistance programs—Bangladesh has more than 100 programs, and India has nearly 950 centrally sponsored schemes, with many more provided at the state level. This plethora of programs usually has more historical or institutional roots than solid technical justification. Some degree of consolidation may be appropriate, but the optimal number of programs is certainly more than one.

Whichever form of social assistance is selected, technology can be harnessed to improve the delivery of social protection programs. In Mexico, geospatial mapping tools are used to identify the most vulnerable areas in cities, down to the block level. Mobile phone data were leveraged to construct poverty maps in Côte d'Ivoire. In Benin, GPS-based data collection located households lacking addresses in urban settlements. Digital technologies can also deliver assistance in fragile places. In Lebanon, electronic smartcards provide 125,000 Syrian refugee households with food vouchers.

Technology is improving the credibility of personal identification systems, which are the first step in the delivery of social protection. In Sub-Saharan Africa, the share of the population with national identifications ranges from nearly 90 percent in Rwanda to less than 10 percent in Nigeria. Technology is also improving access to social registries, which in turn improves coordination among different programs. Better coordination generates cost savings by reducing inclusion errors. In Pakistan, the social registry, which includes 85 percent of the population and serves 70 different programs, contributed to savings of US$248 million. A similar process saved US$157 million in South Africa and US$13 million in Guinea. In Argentina, linking 34 social program databases to the unique identification number of beneficiaries revealed inclusion errors in eligibility for various social programs, resulting in US$143 million in savings over an eight-year period. In 2016 Thailand eliminated 660,000 applicants out of 8.4 million by cross-checking databases using unique national identification numbers.

Payment technologies make a difference. In Ghana's Labor Intensive Public Works scheme, the digitization of paper-based transactions and a wider use of biometric machines reduced overall wage payment time from four months to a week. In the Indian state of Chhattisgarh, the use of electronic devices by the Public Distribution System for food assistance contributed to a reduction in "leakages" from 52 percent in 2005 to 9 percent in 2012.[16]

Social insurance

In June 2011, after six years of double-digit growth, Ethiopia introduced a landmark social insurance law. For the first time, the mandate to provide pension and disability benefits was extended to private sector firms. (However, firms operating beyond the reach of enforcement could evade the law and keep their workers uncovered.) The policy sought to expand social protection and reduce poverty. However, the consequent rise in labor costs, together with other factors, induced firms to adopt more technology. As a result, employment among lower-skill workers dropped, exacerbating the formal-informal divide in the labor market.

The Bismarckian social insurance model of earnings-based contributions is premised on steady wage employment, clear definitions of employers and employees, and a fixed point of retirement. It relies on levying a dedicated tax on wages. In rich countries, this scheme was effective in increasing coverage as workers were steadily absorbed into factories, then into jobs in formal services firms. But this contributory approach is not a good fit for developing countries, where formal and stable employment are not common. Indeed, because eligibility is based on making mandatory contributions, this form of social insurance excludes informal workers, who account for more than two-thirds of the workforce in developing countries and 1 in 10 in India and many countries in Sub-Saharan Africa (figure 6.4). This model is also increasingly unsuitable for the changing nature of work in which traditional employer-employee relationships are no longer the norm. The traditional financing model of social insurance often makes employing workers more expensive, as illustrated by Ethiopia's experience, described above. Rethinking this model is a priority.

FIGURE 6.4 **Coverage of social insurance remains low in most developing countries**

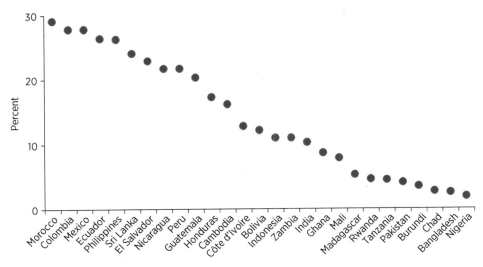

Sources: WDR 2019 team, based on World Bank's pension database and World Development Indicators (database).

A reformed system must ensure that low-income workers have access to effective risk management tools. The right combination of instruments, subsidized for the poorest, is required to cover losses from livelihood disruptions, sickness, disability, and untimely death. Instruments that support stable consumption patterns, or are consumption smoothing, are also important. A comprehensive package of protection in pursuit of these goals would contain, first, a guaranteed minimum insurance with subsidized coverage against impoverishing losses. This instrument would complement social assistance by providing coverage against losses that would be too large to cover through transfers. Second, a mandated savings and insurance plan would allow for consumption smoothing. Finally, market-based "nudged" or purely voluntary savings would allow people to contribute more, if desired. Elements of this model already exist in many countries.

This approach can, along with a guaranteed minimum income, reduce the size and pure-tax element of mandated contributions. To varying degrees, current social insurance models mingle redistribution with risk-sharing functions and require higher contributions that are perceived by many as taxes on work. The extent of redistribution built into current social insurance schemes is low in countries such as Indonesia and Vietnam, but it is substantial in countries such as China and the Philippines. Simulations suggest that a shift like the one proposed here could reduce the payroll tax rate in a country such as the Philippines from 18 to 14 percent.

Some countries are already moving in this direction. The significant extension of the rural pension scheme in China is one example. Currently, around 360 million rural and urban informal workers are contributing to the scheme. Some 150 million older people are receiving payments.[17] Similarly, Costa Rica's government covers part of the pension contribution for the self-employed. Thailand does the same for informal sector workers who choose to join a special pension scheme aimed at low-income workers. Subsidies could be offered to everyone or just to the poor, or they could be gradually reduced as income grows. Turkey's health insurance system does the latter. In addition to providing an almost universal old-age pension, Thailand pays part of the social insurance premium for working-age people in the informal sector. The cost of the subsidy depends on the subsidy level as well as the size of the population to be subsidized.

In many emerging economies, social insurance liabilities are limited because coverage is low. In countries such as Bangladesh, the Lao People's Democratic Republic, Namibia, Somalia, and South Africa, pensions are not financed by labor taxes but by general revenues. In these cases, decoupling from payroll taxes may be feasible. A significant portion could be replaced with other taxes while broadening the coverage beyond those in contracted and regulated standard employment relationships.

Beyond the basic insurance level, additional policy support is likely required to achieve adequate protection. Additional mandated contributions would allow consumption smoothing, but the instruments needed are often missing in countries with underdeveloped capital and insurance markets. This layer would cover formal workers. However, setting the level of insurance is not trivial because a higher mandate leads to higher labor taxes. In some countries, these taxes are already high, which affects formal employment. The average payroll tax rate used to finance contributions is almost 23 percent in advanced economies.[18] It is also more than 20 percent in countries such as China, the Arab Republic of Egypt, and Peru. The mandate could be relaxed by reducing the tax rate or lowering the ceiling on earnings subject to mandatory savings.

To complement mandatory contributions, participation in savings or insurance schemes could be the default option. Some measures include adding an "opt-in" default on business registration and income tax returns. These measures would potentially lower transaction costs. Other approaches that rely on behavioral insights are also instructive in some cases. In Kenya, giving people a golden-colored coin with numbers for each week to keep track of their weekly deposits doubled their savings rate.[19] Another form of nudging may include "commitment devices" through which, for example, people agree to incur a loss if they do not reach a savings goal. Technology vastly increases the possible nudges. Among other things, it facilitates the defaulting of rounding from mobile money or credit card transactions into savings.

Larger national nudge efforts—regardless of the way people work—are also under way to augment savings and insurance efforts. The KiwiSaver program in New Zealand relies on automatic enrollment and offers a limited set of investment choices. The United Kingdom's National Employment Savings Trust, or NEST, operates in a similar way. In both programs, although people are allowed to withdraw savings, incentives dissuade them from doing so.

Labor regulation

In many developing countries, labor regulations were adopted at the time of colonialism. Through conquest, labor law was transplanted throughout Western Europe and the colonies in North and West Africa, Latin America, and parts of Asia. The repercussions from this movement are still being felt in the 21st century: civil law countries have significantly more stringent labor regulations than do common law countries, placing more restrictions on how employers and workers interact.[20]

The more restrictive approach to labor regulation is a poor fit with many developing countries' labor markets because it assumes a greater administrative capacity than most governments have. Designed with industrial-era economies in mind and at a time of weak social protection systems, labor regulations often fail to protect most workers when informality is the norm and work is often out of reach of the authorities. Regulations in most countries are written assuming that most working people are in stable, full-time wage employment. In many developing countries, these types of jobs are an exception, mostly found in the public sector or among high-skill workers.

Reforms would have to address three main challenges associated with labor market regulations. First, these regulations cover only formal workers whose labor is observed by the state. Yet more than half of the global labor force is informal. Second, governments are trying to do too much with labor regulations, expecting them to act as a substitute for social protection, including ensuring a minimum income or substituting for unemployment benefits. And, third, as argued in the *World Development Report 2013: Jobs*, while regulations address labor market imperfections, they often reduce dynamism in the economy by affecting labor market flows and increasing the length of time spent in both employment and unemployment. When regulations are too strict and exclude many workers, especially young and low-skill people, firms find it difficult to adjust the composition of their workforces. The ability to adjust is an important condition for adopting new technologies and increasing productivity.[21]

In a sample of 60 countries, moving from the 20th to the 80th percentile in job security (in countries with strong rule of law) cuts the speed of adjustment to shocks in employment by a third and reduces annual productivity growth by 1 percentage point.[22] The adoption of productivity-enhancing technology is negatively associated with the strictness of some labor regulations, specifically those with burdensome dismissal procedures.[23] Thus the technology-intensive sectors are smaller in countries with stricter employment protection regulations.[24] More stringent regulations are also associated with lower entry and exit of firms—especially small firms—in industries in which labor moves more frequently between jobs.[25] Within countries, similar evidence is also emerging.[26]

To address this challenge, policy makers would have to rethink labor regulations. Some countries are reforming theirs in ways that support firms and workers in adapting to the changing nature of work. Italy's recent reforms have been associated with the creation of more permanent jobs.[27] Aiming for a balance of security and flexibility is vital. Many governments have made their labor markets more flexible. However, only a few are making corresponding investments in income support and reemployment assistance to get workers back into work. Increasing flexibility for firms goes hand in hand with stronger social protection, intermediation and job search assistance programs, and arrangements for strengthening the voice of workers. Beyond basic regulations, protections would be provided to all working people no matter how they engage in the labor market as part of a comprehensive approach to social protection and labor institutions. This approach would provide additional protection to the many workers—often the most vulnerable—who are effectively excluded. This would be a shift away from protecting some jobs to protecting all people.

Reasonable notice periods and protections against discriminatory dismissals are important to counter employer market power. However, when the rules applied to firms' hiring and dismissal decisions are too onerous, they create structural rigidities that carry higher social costs in the face of disruption. Bolivia, Oman, and República Bolivariana de Venezuela do not allow contract termination for economic reasons; they limit grounds for dismissal to disciplinary and personal matters. In 32 countries, an employer needs the approval of a third party even for individual redundancies. In Indonesia, approval from the Industrial Relations Dispute Settlement Board is required for dismissal. In Mexico, an employer must obtain approval from the Conciliation and Arbitration Labor Board. And in Sri Lanka, an employer must obtain the consent of the employee or approval of the commissioner of labor.

Firms could be given more flexibility in managing their human resources contingent on the law mandating proper notice, the presence of an adequate system of income protection, and efficient mechanisms to punish discrimination. However, more flexible dismissal procedures should be balanced with increased protections outside of the work contract and active reemployment support measures to protect people who lose their jobs. Otherwise, reducing restrictions on hiring and dismissal decisions would shift an unmanageable risk burden onto workers. The current approach in many countries, however, places too much of this burden on firms and not enough directly on the state. To reduce the risk of abuse by firms, governments could audit firms based on the associated risk that they will violate the law and then apply penalties to those found at fault.

The provision of financial protection to workers whose livelihood has been disrupted is also ripe for reconsideration. Severance pay is the most prevalent form of this protection in most low- and middle-income economies. However, it is a practice left over from a time when governments were unable to offer unemployment income support. Some countries have, on paper, extremely generous severance pay policies. The statutory severance pay after 10 years of continuous employment is 132 weeks of salary in Sierra Leone, 130 weeks of salary in Mauritius, and 120 weeks of salary in Bahrain.

Yet severance pay is an ineffective instrument for income protection because it pools risk at the firm or industry level, where shocks and losses are correlated. Employees also face a high risk of not receiving payments if their employers have liquidity constraints or go out of business. Placing greater reliance on unemployment benefits organized nationally would give workers more reliable options. National instead of firm-based arrangements would also open this form of protection to all, no matter where or how they work.

To ensure sufficient protection while preserving work incentives, unemployment benefit systems could rely on both individual savings and redistribution. Savings could be drawn on for unemployment or for retraining. If people do not draw down all their savings, the remainder would be available upon retirement. Workers without enough savings would be able to rely on the minimum income guarantee financed through general revenues. Chile and Jordan have individual savings accounts for unemployment. Singapore has individual accounts for housing or education.

Scrutiny of industrial-era employment protections should be accompanied by an assessment of rigid, possibly outdated, laws on work arrangements.

Some new forms of work blur the distinction between being an employee and being a "dependent" self-employed—for example, is a Yandex.Taxi driver in Moscow a Yandex.Taxi employee? To ensure the basic set of protections just discussed, labor codes should define more clearly what it means to be an employee in current labor markets. This definition would be based, for example, on the extent to which workers determine their working conditions (such as when to work). It is important to ensure convergence in the types of benefits and protections that workers receive, regardless of the length of time they spend with an employer.

Finally, strengthening the enforcement of labor laws and mechanisms to expand workers' voices is a worthy goal as well. Moving to a simpler core contract would require stronger collective bargaining structures as fewer protections are prespecified in the law. However, the significance of such structures is declining: across high-income countries the share of workers covered by a collective agreement fell, on average, from 37 percent in 2000 to 32 percent in 2015. Also in 2015, 24 percent of employees were members of trade unions, down from 30 percent in 1985. In developing countries, where there is high informality, unions and collective bargaining tend to play a limited role (figure 6.5). Unionization rates vary between 15 and 20 percent of workers in Brazil, Moldova, Senegal, and Tunisia to less than 10 percent of workers in countries such as Ethiopia, Guatemala, Indonesia, and Turkey.

FIGURE 6.5 Rates of unionization are low and declining in many developing countries

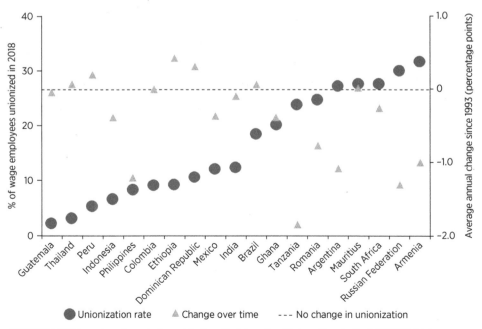

Source: WDR 2019 team, based on select countries from the International Labour Organization's ILOSTAT database.

Digital technologies are also useful in the enforcement of labor laws and mechanisms to strengthen the voice of workers. They bring down enforcement costs by more cheaply monitoring compliance with laws. In Brazil, the Annual Social Information Report is used to monitor compliance with the Apprentice Law. Oman's Worker Protection Scheme allows monitoring of wage payments. And social media plays a role in voicing complaints about employers and working conditions, putting pressure not only on authorities but also on employers because of reputational risks.

To deal with the risks associated with current and future labor markets, governments have to rethink social protection systems. Social assistance should be enhanced, including through a guaranteed social minimum. Potentially, this minimum could be universal, based on countries' conditions and preferences. The notion of progressive universalism may guide any such expansion from the bottom up. Social insurance will play a vital role. Yet the typical Bismarckian model is withering or remains aspirational for many countries, especially because of pervasive informality. As more investments are made in social protection, a balanced approach to labor market regulations could meet productivity and equity goals more effectively.

Notes

1. For a more detailed discussion, see Packard et al. (2018).
2. World Bank (2018).
3. Bastagli et al. (2016).
4. Akee et al. (2018).
5. Ralston, Andrews, and Hsiao (2017).
6. Eurofound (2015).
7. López-Calva and Ortiz-Juárez (2011).
8. Dang and Dabalen (2018).
9. *Economic Times* (2018).
10. Browne and Immervoll (2017).
11. IMF (2017). See Harris et al. (2018) for a discussion on financing a UBI via the value added tax.
12. Ministry of Finance, India (2017).
13. Khosla (2018).
14. Marinescu (2018).
15. Salehi-Isfahani and Mostafavi-Dehzooei (2018).
16. Alderman, Gentilini, and Yemtsov (2017).
17. Dorfman et al. (2013).
18. OECD (2017).
19. Akbas et al. (2016).
20. Botero et al. (2004).
21. World Bank (2012).
22. Caballero et al. (2013).
23. Packard and Montenegro (2017).
24. Bartelsman, Gautier, and De Wind (2016).
25. Bottasso, Conti, and Sulis (2017).
26. Brambilla and Tortarolo (2018).
27. Sestito and Viviano (2016).

References

Akbas, Merve, Dan Ariely, David A. Robalino, and Michael Weber. 2016. "How to Help Poor Informal Workers to Save a Bit: Evidence from a Field Experiment in Kenya." IZA Discussion Paper 10024, Institute of Labor Economics, Bonn, Germany.

Akee, Randall, William Copeland, E. Jane Costello, and Emilia Simeonova. 2018. "How Does Household Income Affect Child Personality Traits and Behaviors?" *American Economic Review* 108 (3): 775–827.

Alderman, Harold, Ugo Gentilini, and Ruslan Yemtsov, eds. 2017. *The 1.5 Billion People Question: Food, Vouchers, or Cash Transfers?* Washington, DC: World Bank.

Bartelsman, Eric, Pieter A. Gautier, and Joris De Wind. 2016. "Employment Protection, Technology Choice, and Worker Allocation." *International Economic Review* 57 (3): 787–826.

Bastagli, Francesca, Jessica Hagen-Zanker, Luke Harman, Valentina Barca, Georgina Sturge, and Tanja Schmidt, with Luca Pellerano. 2016. *Cash Transfers: What Does the Evidence Say? A Rigorous Review of Programme Impact and of the Role of Design and Implementation Features.* London: Overseas Development Institute.

Botero, Juan Carlos, Simeon Djankov, Rafael La Porta, Florencio C. López de Silanes, and Andrei Shleifer. 2004. "The Regulation of Labor." *Quarterly Journal of Economics* 119 (4): 1339–82.

Bottasso, Anna, Maurizio Conti, and Giovanni Sulis. 2017. "Firm Dynamics and Employment Protection: Evidence from Sectoral Data." *Labour Economics* 48 (October): 35–53.

Brambilla, Irene, and Darío Tortarolo. 2018. "Investment in ICT, Productivity, and Labor Demand: The Case of Argentina." Policy Research Working Paper 8325, World Bank, Washington, DC.

Browne, James, and Herwig Immervoll. 2017. "Mechanics of Replacing Benefit Systems with a Basic Income: Comparative Results from a Microsimulation Approach." *Journal of Economic Inequality* 15 (4): 325–44.

Caballero, Ricardo J., Kevin N. Cowan, Eduardo M. R. A. Engel, and Alejandro Micco. 2013. "Effective Labor Regulation and Microeconomic Flexibility." *Journal of Development Economics* 101 (March): 92–104.

Dang, Hai-Anh H., and Andrew L. Dabalen. 2018. "Is Poverty in Africa Mostly Chronic or Transient? Evidence from Synthetic Panel Data." *Journal of Development Studies* (January 8). https://www.tandfonline.com/eprint/82ZSMS2VyytviAfKRkVW/full.

Dorfman, Mark C., Dewen Wang, Philip O'Keefe, and Jie Cheng. 2013. "China's Pension Schemes for Rural and Urban Residents." In *Matching Contributions for Pensions: A Review of International Experience*, edited by Richard Hinz, Robert Holzmann, David Tuesta, and Noriyuki Takayama, 217–42. Washington, DC: World Bank.

Economic Times. 2018. "1 or 2 States May Roll Out Universal Income in Two Yrs: CEA Arvind Subramanian." January 29. https://economictimes.indiatimes.com /news/economy/policy/1-or-2-states-may-roll-out-universal-income-in-two-yrs -cea-arvind-subramanian/articleshow/62696689.cms.

Eurofound (European Foundation for the Improvement of Living and Working Conditions). 2015. "Access to Social Benefits: Reducing Non–Take-Up." Publications Office of the European Union, Luxembourg.

Harris, Tom, David Phillips, Ross Warwick, Maya Goldman, Jon Jellema, Karolina Goraus, and Gabriela Inchauste. 2018. "Redistribution via VAT and Cash Transfers: An Assessment in Four Low and Middle Income Countries." IFS Working Paper 18/11, Institute for Fiscal Studies, London.

IMF (International Monetary Fund). 2017. "Tackling Inequality." *Fiscal Monitor,* World Economic and Financial Surveys, IMF, Washington, DC, October.

Khosla, Saksham. 2018. "India's Universal Basic Income: Bedeviled by the Details." Brief, Carnegie India, New Delhi, February.

López-Calva, Luis F., and Eduardo Ortiz-Juárez. 2011. "A Vulnerability Approach to the Definition of the Middle Class." Policy Research Working Paper 5902, World Bank, Washington, DC.

Marinescu, Ioana. 2018. "No Strings Attached: The Behavioral Effects of U.S. Unconditional Cash Transfer Programs." Roosevelt Institute, New York, May.

Ministry of Finance, India. 2017. *Economic Survey 2016–17.* New Delhi: Oxford University Press.

OECD (Organisation for Economic Co-operation and Development). 2017. *Taxing Wages 2017.* Paris: OECD.

Packard, Truman, Ugo Gentilini, Margaret Grosh, Philip O'Keefe, Robert Palacios, David Robalino, and Indhira Santos. 2018. "On Risk-Sharing Policy for a Diverse and Diversifying World of Work." White Paper of the Social Protection and Jobs Global Practice, World Bank, Washington, DC.

Packard, Truman, and Claudio E. Montenegro. 2017. "Labor Policy and Digital Technology Use: Indicative Evidence from Cross-Country Correlations." Policy Research Working Paper 8221, World Bank, Washington, DC.

Ralston, Laura, Colin Andrews, and Allan Hsiao. 2017. "The Impacts of Safety Nets in Africa: What Are We Learning?" Policy Research Working Paper 8255, World Bank, Washington, DC.

Salehi-Isfahani, Djavad, and Mohammad H. Mostafavi-Dehzooei. 2018. "Cash Transfers and Labor Supply: Evidence from a Large-Scale Program in Iran." *Journal of Development Economics.* In press.

Sestito, Paolo, and Eliana Viviano. 2016. "Hiring Incentives and/or Firing Cost Reduction? Evaluating the Impact of the 2015 Policies on the Italian Labour Market." Questioni di Economia e Finanza (Occasional Paper) 325, Banca d'Italia, Rome, March.

World Bank. 2012. *World Development Report 2013: Jobs.* Washington, DC: World Bank.

———. 2018. *The State of Social Safety Nets 2018.* Washington, DC: World Bank.

CHAPTER 7

Ideas for social inclusion

A social contract envisions the state's obligations to its citizens and what the state expects in return. This basic concept has evolved over time. For much of history, social contracts have been imposed by force or the threat of it. Nowadays, the sustainability of social contracts hinges on how fair they are perceived to be. In his 1762 book *On the Social Contract; or, Principles of Political Rights*, French philosopher Jean-Jacques Rousseau posits that everyone will be free because all forfeit the same number of rights and are subjected to the same duties. This is the view this chapter takes of the social contract: as a policy package that aims to contribute to a fairer society.

Old and new pressures on social contracts are raising calls for new ideas. Cracks in the current social contracts are evident in the lack of efficient public services for most of the poor. Meanwhile, the changing nature of work is generating fears about mass unemployment. These trends are straining the relationships among citizens, firms, and governments across the globe. Although some of these fears appear to be exaggerated, there are indeed reasons to be concerned.

Technological developments in the digital era merit the injection of new ideas into public debates about social inclusion—defined as improving the ability, opportunity, and dignity of those most disadvantaged in society. Two elements deserve special attention. First, using technology, governments have new ways to reach the poor as well as others who lack access to quality services or tools to manage risks. Many work informally in low-productivity jobs without access to protections, making it difficult to escape or remain out of poverty. Informality limits the reach of social insurance systems that are based on formal earnings contributions declared to the state.

Second, the changing nature of work implies adjustment costs for workers. Technology has varying impacts on skills and the demand for them in the labor market. Depending on the technology, some skills (and the workers who possess them) are becoming more relevant than others in the world of work. Advanced skills—such as complex problem-solving and critical thinking—are becoming more valued in labor markets. People with these skills can work more effectively with new technologies. Sociobehavioral skills—such as empathy, teamwork, and conflict resolution—are also becoming more valuable in labor markets because they cannot be easily replicated by machines.

This is the right time to think about how to improve social inclusion. The politics of some of the reforms are complex because of the potential trade-offs between, for example, investments in the current generation of workers versus those in future generations. Public spending has to become more efficient, and additional sources of revenue have to be identified to enhance social inclusion. Aspirations, especially among the youth, are rising, in part due to social media and urbanization. When aspirations are met, they foster opportunity and prosperity. But when aspirations are unfulfilled, they can lead to frustration, or even unrest, in some countries.

This chapter addresses three questions. First, how can society frame a new social contract in the context of high informality and the changing nature of work? Second, if a government is given a mandate to prepare a social contract aimed at improving fairness in society, what would be its basic ingredients? And, third, how can the state finance any proposed reforms? This exercise sets out a scenario for politicians to consider as part of legislative processes and national consultations.

A global "New Deal"

"There is a culture of not participating, of not caring, of silence. The social contract is broken," said a resident in 2017 of one of the areas affected by insecurity in Mexico.[1] Cracks in the current social contracts were already evident in the events surrounding the Arab Spring of 2010–12 and the backlash against globalization that is reflected in rising protectionism. In many developing countries, a dysfunctional social contract may lead citizens to make fewer demands on the state to improve public services. Indeed, some observers have suggested that in developing countries the middle class "send their children to private schools, use private healthcare, dig their own boreholes for water and buy their own generators."[2]

Mechanisms to ensure equal opportunity, which in turn ensures social inclusion, often fall short. Countries are neglecting investments in the early years of children's lives, particularly children in disadvantaged groups. In Latin America, overall per capita government spending on children under age 5 is a third that for children between the ages of 6 and 11. In Sub-Saharan Africa, only 2 percent of countries' education budgets, on average, goes to preprimary education.[3] Tax and social protection systems in developing countries redistribute income to a limited extent. In some countries, this is because taxation lacks progressivity; in many others it is simply because the revenue collected is too low.

Persistently high levels of informality are a symptom of the erosion of social contracts. Informal employment exceeds 70 percent in Sub-Saharan Africa and 60 percent in South Asia. In Latin America, it is more than 50 percent. Informal workers are beyond the reach of the state with respect to the provision of social services, robust social protection, and redistributive measures. People operating in the informal economy evade their obligations to the state by not paying taxes. In some ways, informality reflects a lack of trust in the state.[4]

Recent examples of substantially new social contracts and their elements include Denmark's adoption of "flexicurity," which has its roots in the 19th century. These new social contracts combine labor market flexibility with strong social security and active labor market programs. Other examples include the economic reforms introducing market principles that began in 1978 in China, the Balcerowicz Plan in Poland in 1989, and the Hartz reforms in Germany in 2003. Arguably, however, when people think about social contracts that involve significant reforms associated with the nature of work, the New Deal under U.S. president Franklin D. Roosevelt is a common yardstick for ambition. The reference evokes the possibility of subsidizing employment (or taxing robots) in response to technological progress. The allusion, however, is disingenuous.

During the Great Depression of 1929–33, the U.S. unemployment rate rocketed from 3 percent to 25 percent. Industrial output halved. Responding to the dismal state of the economy, in 1932 Franklin Roosevelt pledged "a new deal for the American people" as he accepted his party's presidential nomination. His New Deal eventually encompassed the various programs and reforms his administration put in place from 1933 to 1938 to lift the United States out of the Depression.

The New Deal, albeit bold and comprehensive, was a response to a problem that is different from that faced in 2018 in the context of informality in developing countries or the changing nature of work globally. Most notably, although the Depression was largely a transitory shock to the U.S. economy, changes to the nature of work and persistent informality are anything but transitory. Some of the measures included in the New Deal—such as the Federal Deposit Insurance Corporation and the Supplemental Nutrition Assistance Program—addressed not only the temporary shock of the Depression but also the permanent need for protections beyond the crisis. However, the largest programs, especially those subsidizing employment or earnings, were temporary—which was appropriate for the circumstances.

Public works activities may go beyond infrastructure. Currently, some essential social activities are being provided voluntarily by individuals. Informal care, or caregiving for a household member with severe disabilities or long-term illness, is widespread. According to recent estimates, more than 2 million people in the United Kingdom are receiving informal care. Women are more likely than men to experience an unexpected move into providing such care. As a result, they face difficulties in balancing their caregiving role with their engagement in the labor market. In addition to lost income, such arrangements can have negative sociobehavioral impacts on the well-being of informal providers.

Effective social care entails reimagining a role for the state in reducing involuntary unemployment by providing services in several areas. These include child care, disability and old-age care, psychological support for the long-term unemployed, support for social kitchens, and rehabilitation from drugs and violence. Interventions such as Kinofelis in Greece and the Expanded Public Works Programme in South Africa are activities of this nature.

An area for enhanced public service provision is community-based primary health care, which extends preventive and curative health services beyond health facilities into communities and households. A review of empirical studies reveals that this approach is effective in enhancing nutrition and immunization, controlling pneumonia and other diseases such as malaria, as well as preventing and treating the human immunodeficiency virus. Equipped with less comprehensive training than professional health workers, community health workers provide basic medical care, have solid referral capabilities, and develop trust in the communities they serve.

Creating a new social contract

Equality of opportunity plays a big role in the changing nature of work. Investing in early childhood development can foster opportunity. One estimate suggests that expansion of early childhood development policies in the United States could reduce inequality by 7 percent and increase intergenerational income mobility by 30 percent.[5] Equality of opportunity also means boosting social protections, including social assistance and insurance, in ways that are compatible with work. These elements of the social contract echo the three freedoms featured by Nobel Prize winner Amartya Sen in *Development as Freedom*: political freedoms and transparency in relations between people, freedom of opportunity, and economic protection from abject poverty.[6]

Beyond some core elements, any new social contract would have to be tailored to its particular country context. One clear area of customization is related to demographic trends. By 2050, more than half of global population growth will have occurred in Sub-Saharan Africa, where the annual growth rates of the working-age population are projected to exceed 2.7 percent.[7] By contrast, the populations of East Asia and the Pacific are aging: more than 211 million people over the age of 65 live in this region, accounting for 36 percent of the global population in this age group. By 2040, the working-age population will have shrunk by 10–15 percent in China, the Republic of Korea, and Thailand.[8] Countries in Sub-Saharan Africa and South Asia would therefore have to be especially responsive to the needs of the large youth cohorts entering the labor market to ensure the sustainability of the social contract. Social contracts in Eastern Europe and East Asia would also need to create mechanisms to finance the protection and care of the elderly in a sustainable manner.

A society with equality of opportunity is often defined as a society that manages to give all its members an equal chance to attain economic and social well-being. However, this happens only if all members of society have access to some guaranteed social minimum, including health care, education, and social protection. Such a minimum would provide basic human capital to everyone, placing them on an equal footing to pursue their goals.

The labor market is increasingly valuing advanced cognitive and sociobehavioral skills that complement technology and make workers more adaptable. This means that inequality will increase unless everyone has a fair shot at acquiring these skills. In fact, in view of the changing nature of work, lack of education is likely to be one of the strongest mechanisms for transmitting inequalities from one generation to the next. A new social contract should seek to level the playing field for acquiring skills. The most direct way to provide fairness is to support early childhood development. Guaranteeing that every child has access to adequate nutrition, health, education, and protection, particularly in the earliest years of life, ensures a solid foundation for developing skills in the future. Because the acquisition of skills is cumulative, the returns to early investments are the highest.

The changing nature of work is turning basic literacy and numeracy into survival skills. They are required to simply navigate life—to buy medication, apply for jobs, and interpret campaign promises. The ability to read and manipulate numbers also serves as a prerequisite for acquiring advanced skills. But for too many children, schooling does not translate into learning. Millions of children in low- and middle-income countries attend school for four or five years without acquiring basic literacy and numeracy. Consequently, guaranteeing access to basic education is not enough.

A social contract for early childhood development would ideally have three components. The first component ensures that children have the essential inputs so they are healthy, well-nourished, and stimulated during their first thousand days (from conception to 24 months of age). This means maternal access to prenatal health care, immunizations, and micronutrients and information for parents on the importance of breastfeeding and early stimulation. The second component ensures access to quality early learning during a child's next thousand days (from 25 to 60 months of age). This means at least one year of quality preschool so that a child is ready for primary school. Preprimary programs should have age-appropriate curricula and qualified teachers. The third component is birth registration, whereby children are recognized by the state and are equipped with the ability to access essential services throughout their lives. Together, all these components—prenatal health care, birth assistance, immunizations, micronutrients, information for parents, quality preschool, and birth registration—make up a basic package that addresses children's early development and learning needs. A more comprehensive package would include investments in safe water and adequate sanitation. Investments to improve air quality are also increasingly important, and research into cost-effective programs is under way.

Some countries are already trying to deliver this type of social contract. In Cuba's early childhood development program, children's growth and development are regularly monitored. At the beginning of each school year, the education sector identifies families who need specific attention. The Chile Crece Contigo program includes a *programa de acompañamiento familiar*, which works with families, pregnant women, and children under grade four who are at social and health risk. Peru has simplified the birth registration process for easier access to early childhood development services. It supports parents in monitoring children's growth and health and engaging in early stimulation activities. France passed a law in 2018 to ensure that all children have access to preschool, starting at age 3.

A social contract on literacy and numeracy would ensure that students master these skills by grade three (approximately by age 10). By this grade, students should be able to read to access the school curriculum. Children who cannot read by grade three struggle to catch up, eventually falling so far behind that no learning occurs at all. The core ingredients of this element would include learning assessments at the end of grade three to identify children at risk, and reading and math assistance for students in grades one to three who need additional support. A more comprehensive package would include ensuring a pupil-teacher ratio of no more than 40:1 in primary grades and providing adequate learning materials with a target of one pupil per textbook in those grades.

Good models for supporting literacy and numeracy by grade three are available, and they are both cost-effective and scalable, even where resources are limited. In Liberia and Malawi, training teachers to better evaluate their students combined with additional materials significantly improved learning in early grades. In Singapore, all students are screened at the onset of grade one. Children who do not attain the appropriate early literacy skills are supported through the Learning Support Programme. These are straightforward approaches. They train teachers to assess their students through ongoing, simple measurements of their abilities to read, write, comprehend, and do basic arithmetic. Children who need additional support receive targeted materials and undertake targeted activities. These models have been tested with success in Ghana, India, Jordan, and Kenya. They serve as a basis for precise design and budget estimates.

The new social contract would also include elements of social protection. The increased risks encountered in the changing nature of work call for adjustments to worker protection. A new social contract could provide a minimum income, combined with basic universal social insurance, that is decoupled from how or where people work. A guaranteed social minimum could take many forms, achieved through a series of programs or by expanding individual interventions. Each of these modalities presents different comparative advantages and has fiscal, political, and administrative implications.

Low- and middle-income countries have made significant headway in social assistance. In Tanzania, spending on conditional cash transfers increased tenfold between 2013 and 2016. The program currently reaches 16 percent of the population and claims 0.3 percent of the gross domestic product (GDP). Spending on conditional cash transfers in the Philippines grew fivefold over 2009–15: the Pantawid program covers 20 percent of the population at a cost of 0.5 percent of GDP. These trends mirror the growth in categorical or age-based programs such as the Child Support Grant in South Africa. The scheme's coverage increased from 1 million beneficiaries in 2001 to 11 million in 2014, absorbing 0.2 and 1.2 percent of GDP, respectively.

Current experiences offer a wide range of tested programs that could be expanded. Whether old or new, programs should share the notion of progressive universalism. This principle deliberately aims at higher levels of coverage, while ensuring that the poor would benefit more and before others. Where exactly in the income distribution one becomes a net beneficiary instead of a net payer is a choice that countries and governments would make themselves.

Social insurance systems that cover old-age and disability pensions are based on a standard employer-employee relationship with limited suitability for developing countries. New forms of work are increasingly challenging this model in advanced economies as well, and, as a result, informal workers often lack access to this kind of support. The system is financed by labor taxes that raise the costs of hiring workers. As social contracts are reimagined, subsidizing a basic level of social insurance—especially for the poor—could be considered. Such a reform could also equalize the costs borne by different factors of production, such as capital and labor, as the financing of the system is at least partly shifted away from labor taxes toward general taxation.

Providing economic opportunities for young adults must be part of social contracts. But the pace of job creation for new labor market entrants has often been slow. For many young people, persistent gaps in access to adequate skills are barriers to employment. International experience with the "productive inclusion" of poor and vulnerable young people reveals that a wide array of programs is available to connect them to wage employment and self-employment. Interventions may include wage subsidies, public works schemes, entrepreneurship grants and asset transfers (often part of "graduation" models), coaching, apprenticeships, internships, and various forms of training. Empirical evidence finds that these programs have mixed effects, with profiling, design, and their particular contexts shaping their cost-effectiveness. For example, wage subsidies may be appropriate in peri-urban contexts with large industrial parks, whereas graduation schemes are largely devised for rural populations (including the transfer of assets such as livestock).

Financing social inclusion

Social inclusion is costly. Simulations suggest that the components of building human capital, including early childhood development and support for literacy and numeracy by grade three, would cost around 2.7 percent of GDP in low-income countries and 1.2 percent of GDP in lower-middle-income countries. The cost of a more comprehensive human capital package is estimated at 11.5 percent of GDP in low-income countries and 2.3 percent of GDP in lower-middle-income countries. These estimates are based on a fully costed model in developing countries, combined with data-driven assumptions. They are the costs of delivering the human capital package, irrespective of income level or coverage of existing programs.

The actual costs could be lower for countries that choose to build on existing programs. Figure 7.1 shows estimates for three scenarios: a low-income country (Mali), a lower-middle-income country (Indonesia), and an upper-middle-income country (Colombia).

How much it costs to provide a guaranteed social minimum would vary according to the context and design choices. A basic social assistance package would cost 9.6 percent of GDP in low-income countries, 5.1 percent in lower-middle-income countries, and 3.5 percent in upper-middle-income countries. These estimates use a universal basic income (UBI), set at the average poverty gap level and aimed at adults. A more ambitious package,

FIGURE 7.1 Low-income countries would pay more than lower-middle-income and upper-middle-income countries for selected elements of a renewed social contract

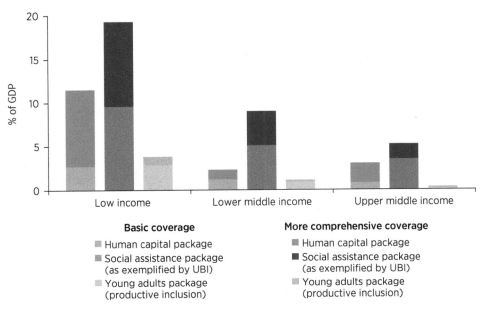

Source: WDR 2019 team. For the human capital package, see Zheng and Sabarwal (2018).

Note: The basic human capital package includes (1) supporting early childhood development, including prenatal health care, birth assistance, immunizations, micronutrients, parental outreach, birth registration, and at least one year of quality preschool for every child; (2) implementing learning assessments at the end of grade three to shine a light on those who are at risk; and (3) offering reading and math assistance for students in grades one to three who need additional support. The more comprehensive human capital package includes, in addition to the basic package, the following elements: (1) access to safe water and adequate sanitation; (2) a pupil-teacher ratio of no more than 40:1 in primary grades; and (3) one pupil per textbook in primary grades. Element-specific unit costs are derived from rigorous studies of relevant in-country programs where available. Alternatively, the most recent cost estimates appropriate for the country's income level are considered. Beneficiary calculations are based on population data from United Nations' World Population Prospects. Other country-level data such as gross domestic product (GDP), access to safe water and sanitation, and prevailing proficiency rates are derived from the World Bank's World Development Indicators (database) and other studies. The basic social assistance package includes universal basic income (UBI) for adults set at the average poverty gap level. The more comprehensive social assistance package includes UBI for the full population set at the average poverty gap level. See chapter 6 for more details on UBI costing. Estimates are based on specific countries for each country group (low income, Mali; lower middle income, Indonesia; upper middle income, Colombia). As such, results are meant to be indicative. See note 11 in this chapter for the young adults estimates method.

illustrated by a UBI that reaches everyone, including children, would cost 9 percent of GDP in lower-middle-income countries and 5.2 percent of GDP in upper-middle-income countries. In the poorest countries, the cost of this package would be in the double digits.[9]

For the 1 billion young adults (20–29 years of age) worldwide, the average intervention costs, depending on the contents of the human capital package, would range from US$831 to US$1,079 per participant.[10] The total cost for reaching vulnerable young adults, or 12.8 percent of the age cohort,[11] would amount to between 2.9 and 3.8 percent of average GDP in low-income countries, from 0.9 to 1.1 percent in lower-middle-income countries, and from 0.2 to 0.3 percent in upper-middle-income countries.

A new social contract would therefore require a significant mobilization of revenue by most governments worldwide. Current taxation patterns reveal

FIGURE 7.2 High-income countries collect a much larger share of their national output in taxes, especially direct taxes, than low-income countries

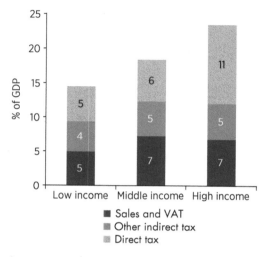

Source: WDR 2019 team, based on the International Centre for Tax and Development's Government Revenue Data Set.

Note: Average values by income group. Data are for 113 countries circa 2015. GDP = gross domestic product; VAT = value added tax.

large differences, especially between low-, middle-, and high-income countries. High-income countries collect a much larger share of their national output in taxes—specifically, direct taxes—than do lower-income countries. Low- and middle-income countries, by contrast, rely more on indirect taxes such as consumption and trade taxes (figure 7.2).

Additional revenue mobilization is possible in most countries. Estimates suggest that Sub-Saharan African countries could raise between 3 and 5 percent of GDP in additional revenue through a combination of reforms that improve efficiency, harness new technologies to improve compliance, and create new sources of taxation.[12]

Governments can reduce tax policy and compliance gaps across a number of fiscal instruments (figure 7.3), including the value added tax, excise tax, and personal and corporate income and property taxes, as well as through fiscal regimes for extractive industries in resource-rich countries.

Often a first line of reform for developing countries, the value added tax is potentially a major source of revenue. But a few countries, such as the Maldives and Myanmar, do not have a value added tax. Many others, particularly in Sub-Saharan Africa, also continue to rely on sales taxes. These countries include Angola, Comoros, Guinea-Bissau, Liberia, and São Tomé and Príncipe. Introducing a value added tax instead of general sales taxes avoids tax cascading (tax paid on tax) by taxing only the value added at each stage of the value chain.

That said, even if a value added tax were in place in emerging economies, it may have only a limited impact on revenue generation. Poor fiscal capacity often results in compliance problems related to flawed implementation. Raising the value added tax thresholds in countries that already have it, closing tax exemptions, and converging toward a uniform tax rate could raise significant revenue, in part by simplifying the system. South Africa and the Sub-Saharan Africa countries of Lesotho, Mauritius, and Senegal do not have many exemptions. By contrast, Cameroon, Malawi, and Zambia have extensive lists of exemptions. In Latin America, in Costa Rica, the Dominican Republic, Honduras, and Uruguay tax expenditures related to the value added tax are estimated at more than 3 percent of GDP in forgone revenue.[13]

In Vietnam, moving to a uniform value added tax rate of 10 percent and significantly narrowing the list of exemptions could increase tax revenues by 11 percent.[14] Informal firms are more likely to pay the value added tax when it is combined with measures to promote payment such as inquiry services, targeted outreach, and incentives that reward compliance. Expanding coverage of the value added tax would also reduce the distortions created between those sectors of the economy that pay the tax and those that do not. Ultimately, such a step would enhance economic productivity and raise revenues further.

The value added tax is often considered regressive relative to income because the poor spend a larger share of their income on consumption than the rich. Even though consumption taxes are regressive when measured as a percentage of household income, they are either proportional or slightly progressive when measured as a percentage of household expenditure. Many countries exempt basic food products such as milk and bread and some medical products from the value added tax to ensure that poor people have access to them at lower costs. Simulations for four low- and middle-income countries—Ethiopia, Ghana, Senegal, and Zambia—show that, although preferential value added tax rates reduce poverty, they do not target poor households effectively. As a result, a UBI funded by 75 percent of the revenue gains from a broader value added tax base—despite being untargeted—could create large net gains for poor households.[15]

Excise taxes are another relatively accessible source of potential revenue. They are simple to implement and are compatible with most tax systems. In 2015 Sub-Saharan African countries collected less than half (at 1.4 percent of GDP) of the excise taxes collected in Europe. There are wide differences

FIGURE 7.3 Especially for low-income countries, the value added tax is a potential resource for financing social inclusion

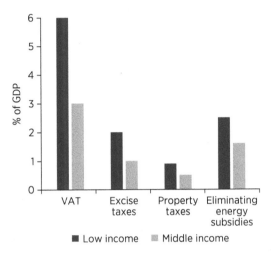

Sources: WDR 2019 team, based on the International Centre for Tax and Development's Government Revenue Data Set; Norregaard (2013); IMF (2015).

Note: For the value added tax (VAT) and excise taxes, estimates are based on the difference between the average tax revenue collected as a percentage of the gross domestic product (GDP) for the top three countries in the income group and the average for all countries in the group. The VAT category includes the value added and sales taxes. For property taxes, the focus is on taxes on immovable property. Estimates for middle-income countries are from Norregaard (2013), who uses a methodology similar to the one followed here for the value added and excise taxes. For low-income countries, where no systematic data are available a conservative estimate of 0.5 percent of GDP is used to reflect the lower capacity of those countries to tax property because complete registries are rare. The potential tax revenue from improved compliance is drawn from IMF (2015), which reports potential gains of about 15 percent, or about 1 percent of GDP, from improved compliance with the value added tax in Latin America. This figure is used as the lower bound for potential gains from improving compliance in the whole tax system. For energy subsidies, estimates are based on the 2015 IMF data set of country-level estimates (IMF 2015). Unlike taxes, resources from eliminating energy subsidies would be available to only those countries that have such subsidies.

in excise tax collection across Sub-Saharan Africa, with several countries, including Benin, Côte d'Ivoire, Madagascar, Mozambique, Nigeria, and Sierra Leone, collecting excise revenues totaling less than 1 percent of GDP.

Excise taxes are often used by governments to achieve social welfare or environmental sustainability objectives by adding in the social cost of negative externalities from the consumption of products such as alcohol, tobacco, and unhealthy foods and from pollution emissions. Some of these taxes are deemed regressive because the poorest families tend to allocate larger shares of their budget to them. This perception should be weighed against the longer-term benefits of these taxes such as lower medical expenses and longer, healthier working lives.

Carbon taxes have become increasingly prevalent. Nationally efficient carbon pricing policies could raise substantial amounts of revenue—estimated at more than 6 percent of GDP in China, the Islamic Republic of Iran, the Russian Federation, and Saudi Arabia.[16] One study of the top 20 carbon dioxide–emitting countries found that, on average, potential revenues raised from nationally efficient carbon pricing would be almost 2 percent of GDP.[17] If revenues from such prices were used to reduce the burden of the broader tax system, the net benefits of carbon pricing could increase substantially. Carbon taxes are currently in place in nearly every large economy except Brazil and the United States, although the rates vary widely.[18] Gradually increasing carbon prices could mitigate the short-term effects on the productive competitiveness of developing economies.

Carbon taxes could be paired with the elimination of energy subsidies for consumption. Globally, government spending on these subsidies amounts to US$333 billion. The fiscal gains from dismantling energy subsidies could be substantial: in many countries their overall level is higher than public spending on social assistance (countries on the right side of the 45° line in figure 7.4). Average spending on energy subsidies in the Middle East and North Africa region is three times higher than on social assistance. Nevertheless, the removal of energy subsidies must undergo a poverty impact analysis, especially for the fuel sources used most intensively by poor households, such as kerosene.

In addition to taxes on goods and services, personal and corporate income taxes can play an important role in increasing revenues in developing countries. Just as technology improves delivery systems for social protection programs, it can facilitate income tax collection by increasing the number of registered taxpayers and social security contributions. The erosion of the corporate tax base affects many countries. It stems mostly from a combination of exemptions (tax incentives) and avoidance loopholes in the international corporate tax system. Higher effective corporate income tax could limit base erosion and profit shifting and address rising corporate market power. Effective tax rates could be increased by streamlining tax expenditures and introducing robust anti–tax avoidance rules such as controlled foreign corporation regimes, limits on interest deductibility, and withholding taxes on payments for services. Withholding taxes are becoming more relevant with the increasing global presence of platform and other firms with a significant digital presence and relatively few tangible assets.

FIGURE 7.4 Some countries spend more on energy subsidies than on social assistance

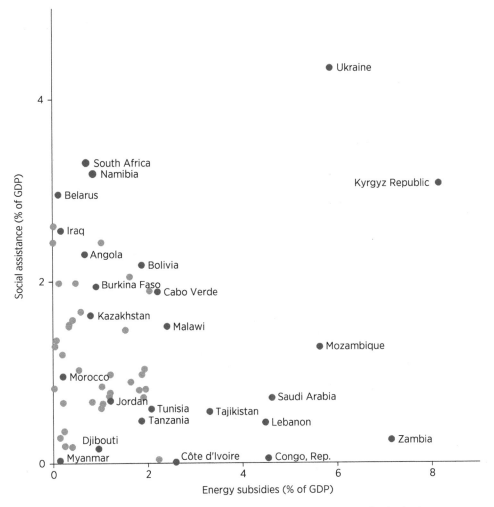

Sources: WDR 2019 team, based on World Bank (2018a) and IMF (2015) database on country-level estimates.

Note: Figure is based on the latest available estimates. GDP = gross domestic product.

Another form of recurrent taxation that can be tapped for further resources in most developing countries is immovable property taxes. These taxes do not distort labor markets, human capital accumulation, or innovation decisions. Property taxes also provide a stable source of revenue that is less susceptible to short-term economic fluctuations and is difficult to evade. And although property taxes would likely not flow into federal social protection schemes (they are typically raised by local governments), they could fund regional or municipal social services or reduce the level of federal transfers to local governments. On average, high-income countries raise 1.1 percent of GDP from immovable property taxes. In middle-income countries, these taxes yield about 0.4 percent of GDP.[19] Yet property taxes represent untapped revenue potential for all countries. This revenue gap is estimated to be 0.9 percent of GDP in middle-income countries and as much as 2.9 percent in high-income countries.[20] Governments in Sub-Saharan Africa are estimated to be missing out on revenues of 0.5 to 1 percent of GDP because of no property taxes whatsoever or their limited application.

Although some countries apply property taxes broadly in the law, the taxes may generate limited revenues because of poor enforcement and informality. Securing broad compliance with property taxes is almost impossible in countries without clear property laws or land cadastres. Formally registered land constitutes less than 1 percent of total land in Cameroon and Rwanda. However, limited registration of land has not prevented the use of specific property taxes in most of Sub-Saharan Africa. For example, property taxes may be levied on leasehold rights, such as in Zambia, or other types of limited property rights, such as concessions in Cameroon and the Democratic Republic of Congo. But even where most property is subject to tax and captured in the register, tax rates may still be too low or the property valuation updated too infrequently to have a meaningful impact on revenues. Although property taxes are becoming more common in Sub-Saharan Africa, some countries, including Botswana, eSwatini (formerly Swaziland), Lesotho, Malawi, and Zimbabwe, still rely on onetime payments. Some countries are taking steps to expand the tax base: Vietnam adopted a tax on nonagricultural land in 2010, and China is considering the imposition of a residential property tax.

Technology can improve property tax collection by digitizing property registration systems. If accompanied by rigorous enforcement, the adoption of new technologies leads to a significant boost in revenues. In 2010 the rate of tax collection on urban immovable property in Lahore, Pakistan, was one of the lowest in the world—0.03 percent of the state's GDP. The average for large cities in developing countries stands at 0.6 percent. The digitization of Lahore's property records in 2012–13 led to the addition of 1.7 million previously unregistered properties. As a result, municipal property tax receipts increased by 102 percent.

Finally, some resource-rich developing countries may be able to raise revenue by introducing or improving regimes applicable to extractive industries. Natural resource taxes and government royalties on oil, gas, and mining could make a substantial contribution to the revenue needs of many emerging economies. The impact of increased production on government revenues has been estimated at about 1 percent of 2011 GDP for Sub-Saharan Africa (assuming a 50 percent government share in rents). The revenue potential is even larger in other countries: 27 percent of GDP in Mozambique from gas exploration and 147 percent in Liberia from iron ore and petroleum exploration.[21]

Preparing for and adapting to the changing nature of work require a strong social contract. While the precise components of such contracts may vary, it is important that they ensure the appropriate investments in education and social protection. Yet sustaining renewed action in these sectors calls for substantial fiscal resources. A range of financing options is available to policy makers, the exploitation of which would require careful technical assessments, combined with political leadership at both the national and global levels.

Notes

1. Fisher and Taub (2017).
2. Desai and Kharas (2017).
3. World Bank (2018b).
4. Saavedra and Tommasi (2007).
5. Daruich (2018).
6. Sen (1999).
7. Canning, Raja, and Yazbeck (2015).
8. Trotsenburg (2015).
9. The level of international poverty line used in the simulations varies by country income category.
10. The maximum cost would include a typical multiprogram graduation package, the cost of which (US$1,079) is calculated as the average cost of six developing country interventions (Banerjee et al. 2015). The lower-cost package is based on vocational training programs, the average cost of which (US$831) is based on the experiences of eight developing countries with such schemes (McKenzie 2017).
11. Because of the lack of youth poverty data, the figure refers to the estimated global unemployment rate among youth in 2016 of 12.8 percent, or 135 million young adults in low- and middle-income countries (O'Higgins 2016).
12. IMF (2018).
13. World Bank (2017a).
14. World Bank (2017b).
15. Harris et al. (2018).
16. Parry, Veung, and Heine (2014).
17. Parry, Veung, and Heine (2014).
18. Djankov (2017).
19. Norregaard (2013).
20. Norregaard (2013).
21. IMF (2012).

References

Banerjee, Abhijit, Esther Duflo, Nathanael Goldberg, Dean Karlan, Robert Osei, William Parienté, Jeremy Shapiro, et al. 2015. "A Multifaceted Program Causes Lasting Progress for the Very Poor: Evidence from Six Countries." *Science*, May 15. http://www.econ.yale.edu/~cru2/pdf/Science-2015-TUP.pdf.

Canning, David, Sangeeta Raja, and Abdo S. Yazbeck. 2015. *Africa's Demographic Transition: Dividend or Disaster?* Africa Development Forum. Washington, DC: World Bank; Paris: Agence Française de Développement.

Daruich, Diego. 2018. "The Macroeconomic Consequences of Early Childhood Development Policies." HCEO Working Paper 2018-010, Human Capital and Economic Opportunity Global Working Group, Economics Research Center, University of Chicago, February.

Desai, Raj M., and Homi Kharas. 2017. "Is a Growing Middle-Class Good for the Poor? Social Policy in a Time of Globalization." Global Economy and Development Working Paper 105, Brookings Institution, Washington, DC, July.

Djankov, Simeon. 2017. "United States Is Outlier in Tax Trends in Advanced and Large Emerging Economies." PIIE Policy Brief 17-29, Peterson Institute for International Economics, Washington, DC, November.

Fisher, Max, and Amanda Taub. 2017. "The Social Contract Is Broken: Inequality Becomes Deadly in Mexico." *New York Times*, September 30. https://www.nytimes.com/2017/09/30/world/americas/mexico-inequality-violence-security.html?_r=0.

Harris, Tom, David Phillips, Ross Warwick, Maya Goldman, Jon Jellema, Karolina Goraus, and Gabriela Inchauste. 2018. "Redistribution via VAT and Cash Transfers: An Assessment in Four Low and Middle Income Countries." IFS Working Paper 18/11, Institute for Fiscal Studies, London.

IMF (International Monetary Fund). 2012. "Fiscal Regimes for Extractive Industries: Design and Implementation." IMF Policy Paper, Fiscal Affairs Department, IMF, Washington, DC, August 16.

————. 2015. "How Large Are Global Energy Subsidies?" IMF Working Paper, IMF, Washington, DC. http://www.imf.org/external/np/fad/subsidies/index.htm.

————. 2018. "Regional Economic Outlook: Sub-Saharan Africa, Domestic Revenue Mobilization and Private Investment." World Economic and Financial Surveys 18, IMF, Washington, DC, April.

McKenzie, David J. 2017. "How Effective Are Active Labor Market Policies in Developing Countries? A Critical Review of Recent Evidence." Policy Research Working Paper 8011, World Bank, Washington, DC.

Norregaard, John. 2013. "Taxing Immovable Property: Revenue Potential and Implementation Challenges." IMF Working Paper WP/13/129, International Monetary Fund, Washington, DC, May 29.

O'Higgins, Niall. 2016. *Rising to the Youth Employment Challenge: The Evidence on Key Policy Issues*. Geneva: International Labour Office.

Parry, Ian W. H., Chandara Veung, and Dirk Heine. 2014. "How Much Carbon Pricing Is in Countries' Own Interests? The Critical Role of Co-benefits." IMF Working Paper WP/14/174, International Monetary Fund, Washington, DC, September 17.

Saavedra, Jaime, and Mariano Tommasi. 2007. "Informality, the State and the Social Contract in Latin America: A Preliminary Exploration." *International Labour Review* 146 (3–4): 279–309.

Sen, Amartya. 1999. *Development as Freedom*. Oxford, U.K.: Oxford University Press.

Trotsenburg, Axel Van. 2015. "How Can Rapidly Aging East Asia Sustain Its Economic Dynamism?" Blog, East Asia and Pacific on the Rise, World Bank, Washington, DC. http://blogs.worldbank.org/eastasiapacific/how-can-rapidly-aging-east-asia-sustain-its-economic-dynamism.

World Bank. 2017a. "Ecuador Development Discussion Notes." World Bank, Washington, DC.

————. 2017b. *Vietnam Public Expenditure Review: Fiscal Policies towards Sustainability, Efficiency, and Equity*. Washington, DC: World Bank.

————. 2018a. *The State of Social Safety Nets 2018*. Washington, DC: World Bank.

————. 2018b. *World Development Report 2018: Learning to Realize Education's Promise*. Washington, DC: World Bank.

Zheng, Yucheng, and Shwetlena Sabarwal. 2018. "How Much Would Expanding Early Childhood Investments Cost?" Unpublished paper, World Bank, Washington, DC.